Freshwater Life in Ireland

*Keys to the more easily identified
Irish freshwater plants and animals
and checklists for most groups*

Cedric S. Woods, M.Sc., Ph.D., M.I.Biol.

Zoology Department
University College Galway

Irish University Press

Le haghaidh na paistí i Mionloch

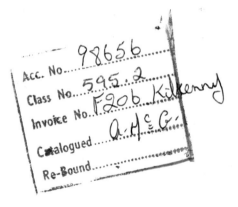
Clothbound ISBN 0 7165 2280 2
Paperback ISBN 0 7165 2281 0

First published 1974 by
Irish University Press Ltd.
81 Merrion Square, Dublin, 2.

Printed in the Republic of Ireland

CONTENTS

FOREWORD

I am grateful for the opportunity to write this foreword. Ireland is fortunate in her abundance and variety of loughs and rivers, and in the fact that, as yet, most of them are relatively unpolluted. Increasingly, however, they are coming under threat of pollution from industrial, domestic and agricultural sources. Often too they are altered by drainage operations. Many fishermen are in no doubt about their deterioration. The conservation of loughs and rivers is probably the foremost environmental problem in Ireland at present. But conservation needs both basic information and public support. I hope that Dr. Woods' book will help to provide them both.

Whether it is called pond-dipping, nature study, natural history, or freshwater ecology, the study of freshwater life is a fascinating and important pursuit. But it is very limited unless the plants or animals found can be named. The Irish flora and fauna are relatively restricted in variety and these keys, uncluttered by species absent from Ireland, will make identification simple. The aesthetic pleasure in looking carefully at plants and animals can be turned to scientific gain once they are identified and recorded. The Irish Biological Records Centre of An Foras Forbartha in Dublin, complemented by similar work at the Ulster Museum in Belfast thrives on such records. Naturalists, between the ages of 8 and 80 who use these keys will become supporters of the conservation ideal and will be able to contribute to its success by collecting information. In this way biological changes in our loughs and rivers, which are more sensitive and informative than the chemical changes that cause them, can be detected. Pollution can be located, seriousness assessed and appropriate action taken. Dr. Woods has performed a service for all of us who like to spend our time "thigh-deep in sedge and marigold".

P.J. NEWBOULD
Professor of Biology,
New University of Ulster.

ACKNOWLEDGEMENTS

My thanks are due to the institutions in which the various drafts of these keys were prepared and used with students – the Wallace High School, Lisburn, and the Department of Zoology and the Board of Extramural Studies, University College, Galway; to the Royal Irish Academy for a grant from the Praeger Fund to collect material and for reference to MS notes of the late Niall MacNeill; to the late Professor Frank Balfour-Browne for the identification of numerous beetles; to IBM Ireland for the loan of an IBM Selectric typewriter for the preparation of camera-ready copy; and to my wife, Doon, for typing and encouragement. The drawings are largely original but many have been adapted freely from a variety of sources, all of which are referred to within the appropriate keys.

4

INTRODUCTION

Keys ask questions which focus attention on one aspect after another of an organism and step by step distinguish it from other species. The observation and inquiry that this process involves is one answer to the question "How does one study life?" The process is stimulating because of the goal of finding a name for the organism. And once a name is found it can be used to delve into the texts and reference books on freshwater biology, working from the index as well as from the table of contents. Or the names can add substance and validity to projects in ecology, pollution monitoring, or conservation. In these ways it is hoped that this small book will open up the study of freshwater life in Ireland to many people who have a latent interest in studying or enjoying to the full the richness of our natural heritage.

The advanced student will find use for the keys to help his memory from time to time but will have more use for the checklists. These give complete lists of the species known to occur in Ireland for each group of plants and animals with the exception of a few large groups of similar species such as mites, water-fleas and midges. Even so, over 600 species are named.

USE OF THE KEYS   Sort your catch into groups of similar specimens as far as possible. Each identification will require answering about five sets of questions starting at the beginning of the appropriate key which can be found from the table of contents on page 3. Read the alternatives 1A, 1B, etc., and decide which set of characters fits your specimen best. Then look to the right opposite this set and use the number shown there to find the next set of alternatives, e.g. 2A, 2B, etc., and so on till a name is reached together with a drawing which matches your specimen.

For example, a plant that has long pointed leaves and floats in the water is traced on page 11, from:-

```
1A      (floating) - - - - - - - - - - - - - - - 2
1B      - - - - - - - - - - - - - - - - - - - - -

   2A      (whole plant floats) - - - - - - - 3
   2B etc - - - - - - - - - - - - - - - - - - - -

3A      - - - - - - - - - - - - - - - - - - - -
3B      (plant large etc.) - - - - - - - - - - - 5

   4A      - - - - - - - -
   4B etc - - - - - - - -

5A      - - - - - - - - - - - - - - - - - - - -
5B      (leaves pointed etc.) = WATER SOLDIER
                                 Stratiotes aloides
```

The language of keys may appear stilted but is, in fact, very precise. Usually the subject comes first and often the verb is understood. For "Leaves absent." (p.15) read "Leaves are absent." and then check on this, or look at your specimen and ask "Are leaves absent?" For "Small; to 10cm long." (p.115) read "Is this specimen about 10cm long, or likely to be when it is fully grown?" If so it may be a minnow and not a young tench or dace.

USE OF THE CHECKLISTS   An identification is final when a species name occurs on its own in a checklist, e.g. *Stratiotes aloides* on p.10 or *Nepa cinerea* (the water-scorpion) on p.82. No other species known to occur in aquatic habitats in Ireland should key out with these independently named species.

Otherwise where a checklist includes a group of names (without gaps between entries) this indicates that these species are all likely to reach the same end-point in the key. For example, the key on p.83 names the genera *Notonecta* (the water-boatman) and *Gerris* (the pond-skater). They have four and seven known Irish species respectively as can be noted easily from p.82. You can now work with and read about these genera knowing how many species are involved. Likewise, where a species name is given in any account, a rapid check can be made to see if it is one known to occur in Ireland. The size of the task is similarly indicated for the advanced student who wishes to identify his material to species level using more detailed keys, but remembering that new Irish records can be expected from time to time.*

Intermediate cases occur, as on pp.48 and 49 where *Physa fontinalis* (the bladder snail) is named in the key, but the checklist shows that one or two other species of *Physa* may be found. The use of the species name (and not just the generic name) in the key implies that the identification is almost certain, in that the other species are uncommon and are only likely to be of interest to the careful specialist.

Common species and ones with striking characters are automatically favoured in the preparation of keys. For this reason the first specimens from any locality to attract your attention have an above average chance of being fully treated by the keys. So practical work will usually get off to a good start. With practice, identification will become easier, if only because each one gives a reference point for comparing related or similar forms. The first *Lymnaea* (pond snail) may prove difficult but by the time that two or three species are familiar the next species will be

*New Irish records by the author are marked "++" and probable new records "+" in the checklists. All other listed species have been recorded in available literature although various avoidable and unavoidable errors may have been included inadvertently.

named more confidently, and more quickly. This will be aided by naming all the snails (or some other group) from any collection at one time. The same result could be served by making named collections of specimens for comparison, or in classwork each of several groups of students could work with a single key for the duration of a project.

The keys have been kept as simple as possible and in most cases they have not used all the recognised diagnostic characters. The author's intention is that they will serve adequately for the novice, and that for more critical use experience is the only sure guide. And experience will be fostered best initially by rapid identifications, then to be tempered by reference to text books and perfected by the use of more detailed keys. Then the experienced worker, aided by his own notes, will find these keys an adequate reminder. The space provided for notes may thus be used for recording additional characters, but may also serve for recording notes on habitat, behaviour, abundance, seasonal availability or other observations. However, distribution may be of most compelling interest and is discussed separately on page 120.

GENERAL REFERENCES TO FAUNA AND FLORA

CLEGG, J. 1956 (2nd edition, 1967). *The Observer's Book of Pond Life*. Warne. London. 209pp.

CLEGG, J. (ed. English edition) 1963 (3rd English edition, revised, 1973). *Blandford Colour Series Pond and Stream Life*. Ysel. Deventer. 108pp.

ENGELHARDT, W. 1964 (2nd revised edition 1973). *The Young Specialist looks at Pond-Life*. Burke. London. 208pp.

ILLIES, J. (Editor) 1967. *Limnofauna Europaea*. Fischer. Stuttgart. 474pp.

MILES, P.M. & H.B. MILES. 1967. *Freshwater Ecology*. (Hulton's Biological Field Studies.) Hulton. London. 126pp.

GENERAL REFERENCES TO FRESHWATER ECOLOGY

BROWN, A.L. 1971. *Ecology of Fresh Water*. Heinemann. London. 129pp.

MACAN, T.T. 1963. *Freshwater Ecology*. Longmans. London. 338pp.

MACAN, T.T. 1973. *Ponds and Lakes*. Allen & Unwin. London. 148pp.

MACAN, T.T. & E.B. WORTHINGTON. 1951. (Revised edition, 1972) *Life in Lakes and Rivers*. Collins. (Fontana New Naturalist) London. 320pp.

POPHAM, E.J. 1955. *Some aspects of Life in Freshwater*. Heinemann. (Scholarship Series in Biology) London. 123pp.

AQUATIC PLANTS

Some knowledge of aquatic plant life is basic to an under-
standing of any aspect of freshwater biology. Physical and chemical differences between lakes, ponds, rivers and streams, and
between one pond and another or one stream and another are reflect-
ed and enhanced by the communities of plants that grow there.
Large stands of plants create particular types of habitats or living conditions for other organisms. Reed beds, or marshy areas
dominated by watercress, or moss-covered stones provide unique
forms of shelter for communities of animals. The minute, but
usually abundant algae are often the major source of food for
animal life, but they are not dealt with here.

A slight change in the physical or chemical conditions of any
habitat would probably lead, slowly, to a change in the dominant
plant type and this would be accompanied by major changes in all
other flora and fauna in the area. Such changes could be due to
the natural evolution of the habitats, to slight climatic changes
or the results of man's changing the surrounding environment.
But before we can understand the subtleties of such changes we
must know the individual plants and animals, their names, relation-
ships, their special adaptations which enable them to live where
they do, and their dependence on each other, as well, of course,
as a knowledge of their present distribution. This is ecology.

So you're off to look at water plants. For the first out-
ing you might choose to look at the dominant plant types in as
wide a range of habitats as possible - canals, rivers, ponds,
swamps or wherever there is permanent water and plants. Such a
survey of your district will probably be of most value if you bring
a map and record the main reed beds. Along muddy shorelines these
will probably be of reeds (*Phragmites communis*), bulrush (*Scirpus
lacustris*), yellow flag (*Iris pseudacorus*), reedmace (*Typha latifolia*), bur-reed (*Sparganium erectum*) and rushes (*Juncus*). Note
the sequence where two or more of these occur close together -
probably, *Scirpus* next to the open water, *Phragmites* in the shallow
water with *Iris* on the marshy banks. How often does *Sparganium*
or *Typha* replace *Phragmites*? What conditions most often coincide
with such changes? Keep walking, and mapping and see if you can
note these patterns; pose questions and answers will come.

Usually, where conditions of water movement allow mud to settle,
the tall plants named above make an unbroken palisade between land
and water. They bend with wind and wave action but die down in
winter to fleshy rootstocks under the mud. Here and there they
are absent. Try and find, by observing conditions generally, if
the gaps are natural or man-made. What other plants have replaced them?

Within, or protected by the reed-beds many other species of

plants form small colonies.   They include horsetails (*Equisetum*), arrowhead (*Sagittaria*), water plantain (*Alisma*) and the beautiful flowering rush (*Butomus* - don't pull it - it's never common). But most other plants belong either in the water or in the marshy area inland from the reeds.   Since far more species are usually involved projects on the smaller plants should be restricted to small areas or to comparisons between two or more well defined areas.

The submerged, or true aquatic plants are in general feathery or leafy when growing in still water (see p.17), streamlined in flowing water (see *Potamogeton* - pond weeds, p.19) or small, tough and reed-like on wave washed shores (see p.21).   Roots of submerged plants are primarily for attachment or may be lacking and the stems and leaves are buoyed up by air inside them.   Most aquatic plants produce flowers, sometimes underwater, but do not rely on seed production for survival from year to year.   So for the winter, portions of them such as special winter buds sink into the mud only to be floated again by the warmth of spring.

Plants with floating leaves, like the reeds, tend to be more permanently fixed in the substrate.   Long rootstocks (stems) have plants growing from buds at intervals and only the rootstocks remain alive over the winter.   Those plants which completely float also form winter buds or else the whole plant sinks when its vital activities slow down with a drop in temperature and light.

The tall reeds provide shelter on their landward side for a host of semi-aquatic species.   Conditions here are optimum for growth since, in addition to shelter from wind and waves, the plants have an assured supply of water, air, sunlight and chemicals. Thus many marsh species have broad leaves and luxuriant growth. Compared to submerged plants they must compete with a greater number of species but need fewer special adaptations in order to survive.   But three groups of plants - bladderworts (*Utricularia*), sundews (*Drosera*) and butterworts (*Pinguicula*) - are very specially adapted to live in those areas of marsh where nitrates are scarce due to bacterial action.   In its own way each obtains this essential chemical by catching and "digesting" animals.

REFERENCES

BURSCHE, E.M.   1971.   *A Handbook of Water Plants*.   Warne. London.   128pp.

MARTIN, W.K.   1972.   *The Concise British Flora in Colour*. Sphere.   London.   336pp.

SCANNELL, M.J.P. & D.M. SYNNOTT.   1972.   *Census Catalogue of the Flora of Ireland*.   Stationery Office.   Dublin.   127pp.

STOKOE, W.J.   n.d.   *The Observer's Book of British Wild Flowers*. Warne.   London.   224pp.

WEBB, D.A.   1943.   (5th. Revised Edition, 1967) *An Irish Flora*. Dundalgan Press.   Dundalk.   259pp.

CHECKLIST   NOTES

LIVERWORTS*
Riccia fluitans
Riccardia sinuata
Ricciocarpus natans

FERNS*
Azolla filiculoides

FLOWERING PLANTS
Lemna minor

L.   *gibba*

L.   *polyrrhiza*

L.   *trisulca*

Hydrocharis morsus-ranae

Stratiotes aloides

*Freshwater, or marginal freshwater, habitats contain a broad range of non-flowering plant types but few species, as follows. Not all of these species are in this key.

ALGAE - Numerous species - planktonic, filamentous, etc.
       - stoneworts - *Chara* spp. and *Nitella* spp. (p.14).
LIVERWORTS - Above.
MOSSES - Five aquatic species (p.14) plus many bog species including *Sphagnum* spp. and *Cinclidotis fontinaloides* which is typical of turloughs.
CLUBMOSSES Etc. - *Lycopodium selago*
              L.   *inundatum*
              Selaginella selaginoides, all from marshy land.
QUILLWORTS - *Isoetes* spp. (p.20) - true aquatics.
HORSETAILS - *Equisetum* spp. (p.22) - two semi-aquatics.
FERNS - *Azolla* sp. (above) - an escape from garden ponds.
         Pilularia sp. (p.20) - a true aquatic.
         Osmunda regalis - the royal fern.
         Thelypteris palustris - the marsh fern.

REFERENCE
     WATSON, E.V. 1968. *British Mosses and Liverworts*. Cambridge University Press. Cambridge. 495pp.

AQUATIC PLANTS

1A   Plants submerged; floating; or submerged with floating leaves.- 2

1B   Most of leaf surfaces above water level; marsh plants.To p.21.-27

    2A   Whole plant floats free at, or just under water surface.   - 3

    2B   Most leaves small and floating with 3-5 lobes.   To p.29. - 40

    2C   Main leaves floating with oval or circular blades. ..... - 6

    2D   Plants totally submerged, and rooted or attached to substrate. Any leaves reaching the surface are similar to submerged ones. (Disregard leaves on flowering.stems.) .... To p.15. -11

3A   Plants small; 1 - 5mm diameter for simple plants, or somewhat larger if many small leaves are present. .... - 4

3B   Plants much larger with leaves rising above the water. ..... - 5

    4A-D   Plants float in surface film. As figured.

A   = WATER FERN - *Azolla filiculoides*

B   =   COMMON DUCKWEED - *Lemna minor*

C   = ... *Lemna gibba*

D = GREAT DUCKWEED - *Lemna polyrrhiza*

    4E-F   Plants float below surface film. As figured.

E   =   IVY DUCKWEED - *Lemna trisulca*

F   = ... Floating LIVERWORTS

5A   Leaf blades almost circular.

FROGBIT - *Hydrocaris morsus-ranae*

5B   Leaves long & pointed.

WATER SOLDIER - *Stratiotes aloides*

11

CHECKLIST   NOTES

Nuphar lutea

Nymphaea alba

Polygonum* amphibium

Potamogeton* coloratus

P.   *natans*

P.   *polygonifolius*

*See p.28 for a marsh species of *Polygonum* and p.18 for submerged species of *Potamogeton*.

6A Leaves almost circular.
    Leaf-stalk joins the blade centrally.   WATERLILIES - 7

6B Leaves oval or pointed.
    Leaf-stalk joins outer margin of blade.   ..... - 8

  7A Leaf-stalks sub-triangular - △ - in section.
      Cabbage-like leaves often present under water.
      Flower yellow; 5cm diameter.
                              YELLOW WATERLILY - *Nuphar lutea*

  7B Leaf-stalks circular in section.
      Cabbage-like leaves rare.
      Flowers white; 10cm diameter. WHITE WATERLILY - *Nymphaea alba*

8A Floating leaves with several
      parallel main veins.   BROAD-LEAVED PONDWEEDS - 9

8B Leaves with one main vein
      with side branches.
                  AMPHIBIOUS BISTORT - *Polygonum amphibium*

  9A Floating leaves thick & tough; all or most broad.
      Submerged leaves narrow.   - 10

  9B Only upper leaves float.
      Leaf shape variable from broad to narrow,
          but on any one plant floating & submerged leaves similar.
      Leaves all thin and almost transparent. *Potamogeton coloratus*

10A Floating leaves 5 - 10cm long; thick. *stipule*
     Submerged leaves with no leaf-blade.
     Stipules (leaf base/sheath) to 5cm.   *Potamogeton natans*

10B Floating leaves narrower & thinner.
     Leaf-blade present on submerged leaves
         but very narrow.
     Stipules to 4cm long.   *Potamogeton polygonifolius*

10C Floating leaves sparse (or absent).
     Submerged leaves thin & elongate; leaf-stalks short (or absent).
     Stipules may be large. .... To p.19. - 22

CHECKLIST   NOTES

STONEWORTS (ALGAE)
    Chara spp.
    Nitella spp.

MOSSES
    Fontinalis antipyretica
    F.   squamosa
    Eurhynchium riparioides
    Drepanocladus fluitans
    D.   aduncus

FLOWERING PLANTS
    Hottonia palustris

    Elodea canadensis

WHORL — a group of several leaves or branches which arise as
        a ring from around a stem.

REFERENCE
    See p.10 for mosses.

11A   Leaves entire (= undivided). .... - 12

11B   Leaves, at least lower ones, divided. .... - 13

11C   Leaves absent. Stem may have whorls of side-branches. - 14

   12A   Leaves small & simple; in 3 series & pointed,
         or with whorls of 3 leaves. .... - 15

   12B   Leaves simple; in pairs, or in whorls of about 10 leaves. - 16

   12C   Leaves come off stem singly (occasionally paired).
         Leaves usually over 30mm long; grass-like,
         or narrower; sometimes broader. .... PONDWEEDS - 19

   12D   Leaves arise from underground stem;
         often all grouped together. .... To p.21. - 24

13A   Leaves in regular whorls of 4 to 8 leaves. .... - 17

13B   Leaves come off stem singly, or irregularly. .... - 18

13C   Flower stem arises from a rosette of
      comb-like leaves and branches.
      Leaves about 50-80mm long.   WATER VIOLET - *Hottonia palustris*

   14A   Stems thin - all under 1mm thick.
         Side branches in whorls; may branch
            again to form a dense, or a straggling plant.
         Rough to the touch; brittle; & has a characteristic odour.
                       STONEWORT - *Chara* or *Nitella*

   14B   Main stem and branches segmented & ribbed. To p.23.
         Branches in whorls; usually above water. HORSETAILS - 28

15A   Stem almost covered by leaves which arise spirally.
      Dark green. Leaves to 5mm long; pointed.
      Usually in clumps on stones.
                    WILLOW MOSS - *Fontinalis antipyretica*

15B   Leaves in whorls of 3. Dark green.
      Leaves to 10mm long.
      May have long roots in mid-water.
      Stem branches irregularly. CANADIAN PONDWEED - *Elodea canadensis*

CHECKLIST   NOTES

Elatine hexandra
E.   hydropiper
Naias flexilis

Callitriche hamulata
C.   hermaphroditica
C.   obtusangula
C.   platycarpa
C.   stagnalis
C.   truncata

Myriophyllum alterniflorum
M.   spicatum
M.   verticillatum

Ceratophyllum demersum

Utricularia intermedia
U.   minor
U.   neglecta
U.   vulgaris

Ranunculus* aquatilis
R.   baudotti
R.   circinatus
R.   fluitans
R.   penicillatus
R.   trichophyllus

*See p.28 for marsh species of *Ranunculus*.

16A   Leaves to 15mm long; in pairs on short
         stems; blades rounded on short stalks.
        Low creeping growth.

WATERWORT - *Elatine*

16B   Leaves to 10mm long; in pairs on long
         stems; blades elongate without stalks.
        Usually forms bushy plants.

STARWORT - *Callitriche*

16C   Leaves larger; to 25mm; in whorls of 8 to 10.
        Stem thick; does not branch.
        Grows straggly under water; head
         may grow erect above water.

MARE'S TAIL - *Hippuris* To p.23. - 28

17A   Three, 4 or 5 leaves in each whorl.
        A leaf consists of a main rib and
         10 - 30 side veins; no blade.
        No insect traps near base (18A). WATER MILFOIL - *Myriophyllum*

17B   About 8 leaves in each whorl.
        Each leaf divides 2 or 3 times
         to give 4 - 8 ends.
        Leaves to 20mm long; slightly spiny.
        Long stems, loosely branched.   - HORNWORT -
        No insect traps near base (18A).   *Ceratophyllum demersum*

18A   Leaves finely divided; irregularly
         spaced on stems.
        Small insect traps present at bases
         of leaves or on separate stems.   BLADDERWORT - *Utricularia*

18B   All leaves, or most, finely divided.
        Leaves come off stem singly.
        Form of plant highly variable.

WATER CROWFOOT - *Ranunculus*

19A   Leaves grass-like or narrower
         arise singly (in pairs in two species).   - 20

19B   Leaves mostly intermediate in breadth between A & C.   ....- 22

19C   Some, or most, leaves broad - their
         length less than twice their breadth.   ..... - 23

CHECKLIST   NOTES

Zannichellia palustris

Potamogeton* pectinatus
P.   filiformis

P.   berchtoldii
P.   pusillus

P.   crispus

P.   obtusifolius
P.   friesii

P.   gramineus
P.   alpinus

P.   lucens

P.   perfoliatus

P.   praelongus

Groenlandia densa

*See also p.12 for species of *Potamogeton* with floating leaves.

20A   A true grass with nodes & ligules   *ligule*
       where leaves come off stem.

To p.23. - 30

20B   Not a grass.   ....   *node*

                          ....   - 21

    21A   Leaves come off stem in pairs or threes;
         hair-like, 30 - 70mm long.
         Stipules (sheaths at base of leaves)
         small.   HORNED PONDWEED   *Zannichellia palustris*

    21B   Leaves come off singly;
         hair-like with a single main vein;
         50 - 100mm long; stipules long.
               Potamogeton pectinatus or *P. filiformis*

    21C   Leaves alternate as in B;
         narrow with 3 main veins.
         Stipules short or absent.
         SMALL PONDWEED - *Potamogeton berchtoldii* or *P. pusillus*

22A   Leaves about 50 -100mm long; to 12mm broad;
       margins wavy & finely toothed.
    Stipules small or absent.
    Plant often reddish in colour. CURLY PONDWEED-*Potamogeton crispus*

22B   Stem slightly flattened.
    Leaves grass-like - 3 to 5 veins; rounded ends.
    Stipules short but broad.
              Potamogeton obtusifolius or *P. friesii*

22C   May have broad floating leaves.
    Submerged leaves narrow & tapering;
       60 - 150mm long; to 6mm broad.
    Stipules medium to large.   *Potamogeton gramineus* or *P. alpinus*

    23A   Leaves (& stipules) large;
         to 100 by 50mm; semi-transparent; wavy.
         Cross veins show clearly. SHINING PONDWEED - *Potamogeton lucens*

    23B   Leaves firmer & clasping the stem.
         Stipules medium sized or absent,
         respectively.   *Potamogeton praelongus* or *P. perfoliatus*

    23C   Leaves in pairs without leaf stalks.
         Stipules absent.   *Groenlandia densa*

CHECKLIST   NOTES

SPORE-BEARING PLANTS
Isoetes lacustris
I.   echinospora

Pilularia globulifera

FLOWERING PLANTS
Subularia aquatica

Littorella uniflora

Rhynchospora fusca
R.   alba

Schoenus nigricans

Triglochin palustris

Lobelia dortmanna

Eriocaulon septangulare

Limosella aquatica

24A   All leaves (or leaf-like stems) cylindrical
      and arising from ground level. .... -▨- ..... - 25

24B   Leaves flattened.   If narrow, then grass-like. .... - 26

25A-D Usually small; under 20cm high.   (If taller see 32.)

A = QUILLWORT - *Isoetes*

A-

B-

B = PILLWORT - *Pilularia globulifera*

C-

C = AWLWORT - *Subularia aquatica*

D-

D = SHORE-WEED - *Littorella uniflora*

26A-C Leaves narrow.   Stems cylindrical (if triangular, see 33, p.25).

A = BEAK-SEDGE -
                     Rhynchospora

26D-F Leaves broader.

B = BLACK BOG-RUSH -
                     Schoenus nigricans

C = ARROW-GRASS -
                     Triglochin palustris

A-

E-

D = WATER LOBELIA -
                     Lobelia dortmanna

B-

D-

E = PIPEWORT -
                     Eriocaulon septangulare

C-

F-

F = MUDWORT -
                     Limosella aquatica

27A   Erect plant with single stem 2-5mm thick, & either with whorls of
      small leaves, or, with segments & may have fine branches. - 28

27B   Leaves elongate; blades simple.   Stem may be leaf-like or v.v.
      Flower stem from base of plant.   GRASSES, RUSHES, REEDS, etc.- 29

27C   Leaf blades undivided; on long stalks from ground.   To p.25. - 35

27D   Leafy plant; blades may be divided. Often branching.To p.27. - 36

CHECKLIST   NOTES

HORSETAILS
Equisetum palustre

E.   *fluviatile*

FLOWERING PLANTS
Hippuris vulgaris

Phragmites communis

Phalaris arundinacea

Glyceria maxima

G.   *fluitans*
G.   *declinata*
G.   *plicata*

Scirpus lacustris
S.   *tabernaemontani*

Eleocharis acicularis
E.   *multicaulis*
E.   *palustris*
E.   *quinqueflora*
E.   *uniglumis*

Juncus acutiflorus
J.   *articulatus*
J.   *bulbosus*
J.   *effusus*
J.   *subnodulosus*

Additional species, which may key out at no. 30 (grasses) are –
Alopecurus geniculatus; *Calamagrostis stricta*; *Catabrosa aquatica*;
at 32 – *Scirpus caespitosus*; *S. fluitans*; *S. cernuus*; *S. maritimus*
and *S. sylvaticus*, also *Blysmus rufus*.

28A   Leaves flat; numerous.
Stem not hollow.
Light green colour.

A =
MARE'S TAIL -
Hippuris vulgaris

28B   Stem, and branches, in
sections; hollow; with
about 6 teeth & ribs
To 30cm high.

B = MARSH HORSETAIL -
Equisetum palustre

28C   As B, but larger; about 12 teeth & ribs.

C = WATER HORSETAIL -
Equisetum fluviatile

29A   Leaves flat, with veins; alternate from stem. ... GRASSES- 30

29B   Stems &/or leaves mostly erect from base. REEDS and RUSHES- 31

30A   Stem unbranched; tall, to 250cm.
Leaves to 10mm wide.
Flower heads dark &
feathery. Ligule hairy.

A =
THE REED -
Phragmites communis

30B   As for small *Phragmites*, but
ligule large; flowers light.

B =
REED CANARY GRASS -
Phalaris arundinacea

30C   Similar again; to 175cm tall; leaves broader.
Flower heads not feathery. Ligule shorter.

C =
REED-GRASS -
Glyceria maxima

30D   Leaves mostly floating; about 6mm
broad by 250mm long. Ligule triangular.

Possibly FLOTE-GRASS -
Glyceria fluitans

31A   Stems &/or leaves cylindrical in section; not triangular. - 32

31B   Stems &/or leaves triangular in section and wiry. ...... - 33

31C   Not as A or B; leaves usually broader, flatter & fleshy. - 34

32A   Stems tall; 120 - 300cm (-10ft).
Flower heads branched.

A = SEDGE or BULRUSH-
Scirpus lacustris

32B   Stems 5 - 45cm long; no leaves.
Flower heads compact.

B = SPIKE-RUSHES -
Eleocharis

32C   Stems 15 - 75cm long; few leaves.
Flower heads loosely branched.

C = RUSHES - *Juncus*
(If smaller see 24)

CHECKLIST   NOTES

Carex spp.
Cladium mariscus

Eriophorum angustifolium
E.   latifolium
E.   vaginatum
E.   gracile

Butomus umbellatus

Typha latifolia
T.   angustifolia

Iris pseudacorus

Sparganium erectum
S.   angustifolium
S.   emersum
S.   minimum

Dactylorhiza incarnata
D.   majalis

Sagittaria sagittifolia

Alisma plantago-aquatica
A.   lanceolatum
Baldellia ranunculoides

Drosera rotundifolia
D.   anglica
D.   intermedia

Hydrocotyle vulgaris

Caltha palustris

Additional species, which may key out at nos. 33 are – *Scirpus triquetrus*, and nearly 30 species of sedge (*Carex*); at 34 – *Acorus calamus; Spiranthes romanozoffiana; Epipactis palustris; Hammarbya paludosa; Ophrys insectifera* and *Narthecium ossifragum;* and at no. 35 – *Viola palustris; V. canina* and *Parnassia palustris.*

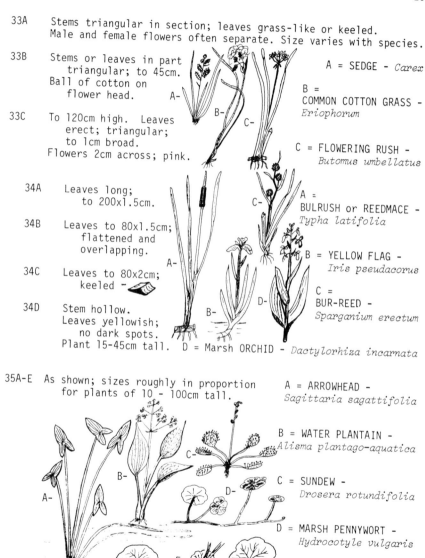

33A   Stems triangular in section; leaves grass-like or keeled.
      Male and female flowers often separate. Size varies with species.

33B   Stems or leaves in part
         triangular; to 45cm.
      Ball of cotton on
         flower head.   A-

33C   To 120cm high. Leaves
         erect; triangular;
         to 1cm broad.
      Flowers 2cm across; pink.

A = SEDGE - *Carex*

B =
COMMON COTTON GRASS -
Eriophorum

C = FLOWERING RUSH -
      Butomus umbellatus

   34A   Leaves long;
            to 200x1.5cm.

   34B   Leaves to 80x1.5cm;
            flattened and
            overlapping.
                            A-

   34C   Leaves to 80x2cm;
            keeled -

   34D   Stem hollow.
         Leaves yellowish;
            no dark spots.
         Plant 15-45cm tall.

A =
BULRUSH or REEDMACE -
Typha latifolia

B = YELLOW FLAG -
      Iris pseudacorus

C =
BUR-REED -
Sparganium erectum

D = Marsh ORCHID - *Dactylorhiza incarnata*

35A-E   As shown; sizes roughly in proportion
           for plants of 10 - 100cm tall.

A = ARROWHEAD -
      Sagittaria sagattifolia

B = WATER PLANTAIN -
Alisma plantago-aquatica

C = SUNDEW -
      Drosera rotundifolia

D = MARSH PENNYWORT -
      Hydrocotyle vulgaris

E = MARSH MARIGOLD -
      Caltha palustris

CHECKLIST   NOTES

Mentha aquatica
M.   spp.

Lycopus europaeus

Teucrium scordium

Stachys palustris
Scutellaria galericulata
S.   *minor*

Lythrum salicaria

Veronica beccabunga

V.   *scutellata*

Epilobium* palustre

Bidens cernua

Additional species which may key out at no. 38 are - 6 species
of *Mentha*; *Ajuga reptans*; *Scrophularia aquatica* and *S. umbrosa*; &
at no. 39 - *Sagina nodosa*; *Anagallis tenella*; *Montia fontana*;
Lythrum portula; *Stellaria palustris*; *S. alsine*; *Veronica
catenata*; *V. anagallis-aquatica*; *Epilobium angustifolium*; *E.
parviflorum*; *Hypericum canadense*; *H. elodes*; *H. hirsutum*; *H.
tetrapterum*; *Lysimachia vulgaris* and *Mimulus guttatus*.

*See p. 28 for other species of *Epilobium*.

36A   Leaves in opposite pairs.   .... - 37

36B   Leaves come off stem singly or from the base in a rosette. - 40

37A   Stem square in section.   .... - 38

37B   Stem round in section.   .... - 39

38A   To 70cm high; bushy.
       Flower head compact.
       Mint smell & taste.   A = WATER MINT -
                                                              Mentha aquatica
38B   Tall; to 80cm high.
       Flowers at base of leaves.
       No aromatic smell.   B = GIPSYWORT -
                                                             Lycopus europaeus
38C   Small; creeping.

38D   Tall; to 80cm high.   C = WATER GERMANDER -
       Stem hollow.   Teucrium scordium

                                                         D = MARSH WOUNDWORT -
                                                             Stachys palustris

39A   To 90cm high.
       Leaves 5-10cm;   A = PURPLE LOOSESTRIFE -
           light green.   Lythrum salicaria

39B   Thick hollow
           stem;   B = BROOKLIME -
           creeping.   Veronica beccabunga

39C   To 45cm high.
       Semi-erect.
       Leaves 2-5cm.   C = MARSH SPEEDWELL -
                                                             Veronica scutellata

39D   To 60cm high.
       Branched.   D = BOG WILLOW-HERB -
                                                             Epilobium palustre

39E   To 60cm high.
       Stem stout;   E = NODDING BUR-MARIGOLD -
       Branched to give bushy plant.   Bidens cernua

CHECKLIST   NOTES

Menyanthes trifoliata

Rorippa nasturtium-aquaticum
R.   *microphylla*
R.   *amphibia*
R.   *islandica*
R.   *sylvestris*
R.   *palustris*

Filipendula ulmaria

Apium nodiflorum
A.   *inundatum*

Ranunculus* omiophyllus
R.   *hederaceus*
R.   *tripartitus*

R.   *flammula*
R.   *lingua*

Polygonum* hydropiper
P.   *mite*
P.   *minus*

Myosotis caespitosa
M.   *secunda*
M.   *scorpioides*

Epilobium hirsutum
E.   *obscurum*

Pinguicula grandiflora
P.   *lusitanica*
P.   *vulgaris*

Additional species which may key out at no. 43 are- *Thalictrum flavum; Trollius europaeus; Potentilla palustris; Sanguisorba officinalis; Cicuta virosa; Sium latifolium; Berula erecta; Oenanthe fistulosa; O. aquatica; Eupatorium cannabinum; Bidens tripartita; Pedicularis palustris;* and at no. 44 - *Cirsium palustre; Crepis paludosa; Taraxacum paludosum; Samolus valerandi* and *Rumex hydrolapathum.*

*See p.16 for submerged species of *Ranunculus*, p.12 for a species of *Polygonum* with floating leaves and p.26 for *Epilobium* spp.

40A   Leaves divided into leaflets or deeply indented.   ....   - 41

40B   Leaves not so; may be toothed.   ....   - 44

41A   Leaf margins smooth.   ....   - 42

41B   Leaf margins toothed.   ....   - 43

42A   Leaves arise singly;
        3 leaflets each
        to 5cm long.

A = BOGBEAN -
Menyanthes trifoliata

42B   Five or more
          leaflets.
        Creeping habit.

A-

B-

B = WATER-CRESS -
Rorippa

43A   Erect stem; to 100cm tall.
        Many irregular
          leaflets.

A-

A = MEADOW SWEET -
Filipendula ulmaria

43B   Stem thick, hollow
        & soft. Leaves
        to 20cm long.

B-

B = WATER-CELERY -
Apium

43C   Leaves small on
        long stalks; floating &
        incompletely divided; lower ones finely divided.

C-

C = WATER CROWFOOT -
Ranunculus

44A   To about 60cm
        high; leaves
        to 20cm long.

A-

A = LESSER SPEARWORT -
Ranunculus flammula

44B   To 50cm tall;
          branched.
        "Hot" to taste.

B-

B = WATER PEPPER -
Polygonum hydropiper

44C   Creeping habit.

C-

C = FORGET-ME-NOT -
Myosotis

44D   Erect; 30-100cm
          high.
        Branched or not.

D-

D = WILLOW-HERB -
Epilobium

44E   Leaf margins incurved.
        Leaves in basal rosettes; sticky.

E-

E = BUTTERWORT -
Pinguicula

WORMS AND SIMPLE ANIMALS

Few areas of permanent freshwater lack sponges, hydras, bryozoans, flatworms, true worms, or leeches, yet these are easily overlooked. The simplest way to ensure their capture is to sample using a combination of the following basic techniques.

On starting to catch animals in any locality look at the upper surface of stones or other solid objects. These are usually not well colonised but you may see some of the black or brown flatworms; and occasionally one of the parasitic leeches waiting for the chance to attach itself onto a bird in the case of *Theromyzon* or a fish in the case of *Piscicola*. Then turn each stone and first look in the depression that it leaves and perhaps collect some of the mud or gravel under it, unless this is compacted. Later, some of the small worms may be isolated from this material.

Quickly look at the sides and undersurface of the stone. Many of the simpler animals shun full daylight, so expect to find flatworms, sponges, bryozoans and leeches here. At first you will require practice to see some of the smaller forms; for example, bryozoans may be missed if leeches have caught your attention. For this reason it may be good policy to search for the smaller, or rarer forms separately. These may still evade you unless you bring samples of weed etc. back with you for more leisurely examination under good illumination. For example, hydras and bryozoans will contract instantly they are disturbed and may take several minutes or up to an hour or so to open out again. Once you have isolated them and observed their behaviour you will have a better chance of recognising them in field conditions.

These techniques will be effective for snails but are not as useful for the mud-dwelling mussels or the more active crustaceans, insects or fishes. Nets are the answer, but the flimsy kind sold in shops is of limited use. Ideally a variety of nets is needed, each for special duty under various conditions. However, for a start, a solid frame, either oblong or triangular, is needed and the fabric should be fine but strong. Such a net can be used to bring up mud which can then be washed out to leave the animals for inspection; it can be pushed into the gravel of a stream and held while animals are washed into it by the current; it may be thrust forcibly into weed to dislodge various species; or it might be trailed in the water to catch floating forms.

In these ways you are sampling - that is, taking a cross-section of the forms of life from a locality. As your interest grows you can refine this process to get estimates of the relative or absolute abundance of species. But it is difficult to get satisfactory results and experts often find difficulty in

convincing themselves of the validity of their own or others'
figures.

A useful quantitative method for estimating the abundance
of sponges, flatworms, leeches and snails, assuming you need to
work only in shallow water in a uniform area, is to record the
number of specimens obtained in a set time interval such as half
an hour.   Your results can be used to compare densities between
one locality and the next.

Comparative work of this sort is often revealing especially
when you have a set purpose.   If pollution of some sort is
suspected, or if you wish to evaluate the effect on stream life
of a known source of pollution then compare the animals and maybe
the plants from several stations.   One or two stations should
be above the source, one or two in areas which are obviously
affected and one at least should be further downstream where
recovery is likely to be almost complete.

Your job is then to sample each station thoroughly and to
record the results uniformly, preferably in a table with a list
of species cross-referenced by the sampling stations.   Thus
you might get:-

Species	Station	(A = upstream; Pollution source A-B)				
	A	B	C	D	E	F
Sponges	12	0	0	0	2	5
Dendrocoelum	6	0	8	47	19	5
Erpobdella	0	0	24	17	3	6

You might then reasonably infer that the pollution adversely
affects sponges; that *Dendrocoelum* cannot tolerate it in high
concentration but somehow benefits from it in low concentration;
and that this applies similarly to *Erpobdella*.   Perhaps the
pollution is organic and is a food for *Asellus* and tubificid
worms which are then prey to *Dendrocoelum* and *Erpobdella*
respectively.   Thus the more species you record the more valid
and less speculative your results will be.

CHECKLIST   NOTES

SPONGES
Ephydatia fluviatilis
E.   mulleri
Heteromeyenia ryderi
Spongilla fragilis
S.   lacustris

HYDRAS**
Hydra attenuata
*H.   circumcincta**
*H.   oligactis**
H.   viridissima

BRYOZOANS
Cristatella mucedo

Plumatella repens
P.   fruticosa
Fredericella sultana
Paludicella articulata

Lophopus crystallinus

*Not definitely known for Ireland.

**Hydra has two relatives in freshwater:-
*Cordylophora lacustris** forms a branching colony (cf. Bryozoans) &
Craspedacusta sowerbii[+] is a minute hydroid with a jellyfish stage.

REFERENCES

ALLMAN, G.J. 1856. *A monograph of the Fresh-water Polyzoa, including all the known species both British and Foreign.* Ray Soc. London. 119pp. + plates.

GRAYSON, R.F. 1971. The Freshwater Hydras of Europe. 1. A Review of the European species. *Arch. Hydrobiol.* 68(3): 436-449.

GRAYSON, R.F. & D.A. HAYES. 1968. The British Freshwater Hydras. *Country-side* 20 n.5. 12: 539-546.

NICHOLS, A.R. 1912. Polyzoa. 2. Freshwater Polyzoa. Clare Island Survey. *Proc. Roy. Irish Acad.* 31(53): 12-14.

STEPHENS, J. 1920. The Freshwater Sponges of Ireland. *Proc. Roy. Irish Acad.* 35 B 11: 205-254.

WORMS AND SIMPLE ANIMALS

1A   Worm-like; or shape irregular, even plant-like.
       Often flattened; may be branched, individuals living as a colony.
       May have segmented body, or bristles, or suckers or eyes.
       Never with a shell or legs or fewer than 12 segments. ...... - 2

1B   Not as above.   Try another key, p.3.   -

     2A   Shape irregular, branched or plant-like.
           Fixed immovably or semi-permanently to a plant, stone, etc.- 3

     2B   Bi-laterally symmetrical; worm-like; rounded or flattened.
           Moves by gliding, looping, swimming or in a worm-like manner.
           May have eyes or suckers and may be segmented. .... - 5

3A   Encrusted mass of irregular shape
          and bread-like texture.
       Usually whitish or light in colour.
       To several cms diameter.   FRESHWATER SPONGE

3B   Small; to about 15mm; fixed to substrate.
       With long tentacles at free end.
       May be branched.   Highly contractile.
       Bright green or gray-brown.   HYDRA

3C   Colony of many small animals;
          compact or branched.
       Each individual has a U-shaped "head"
          fringed with numerous short tentacles.   POLYZOA = BRYOZOA - 4

     4A   Colony oval; to 50mm long.
           Compact; up to 200 "heads".
           Capable of slow locomotion (5mm/hour).   *Cristatella mucedo*

     4B   Colony much branched;
           to several cms across.
           Fixed to hard surface.
           Numerous "heads" on long stalks.   *Plumatella repens*

     4C   As for B but smaller;
           to 10mm across.
           About 12 "heads" at most.   *Lophopus crystallinus*

CHECKLIST   NOTES

NAIDID WORMS

Chaetogaster cristallinus
C.   diaphanus *
C.   diastrophus
C.   limnaei *
Nais barbata
N.   communis
N.   elinguis
Ophidonais serpentina
Stylaria lacustris
Vejovskyella comata

TUBIFICID WORMS

Aulodrilus pluriseta
Branchiura sowerbyi
Ilyodrilus templetoni
Limnodrilus claparedeianus
L.   hoffmeisteri
L.   udekemianus
Potamothrix bavaricus
P.   hammoniensis
P.   moldaviensis *
Peloscolex ferox
Psammoryctides barbatus
Rhyacodrilus coccineus
Tubifex ignotus
T.   tubifex

OTHER WORMS

Eiseniella tetraedra

Lumbriculus variegatus

Gordius violaceus

Haplotaxis gordioides **

Nematodes

*Recent finds by D.J. McGrath (*in lit.*). **Not definitely recorded.

REFERENCES

BRINKHURST, R.O. 1971. *A guide for the identification of British aquatic oligochaeta.* Freshwater Biol. Assn. Sci. Publ. 22. 55
KENNEDY, C.R. 1964. Studies on the Irish Tubificidae. *Proc. Roy. Irish Acad.* 63 B (13): 225-237.

5A   Simple, unsegmented worms; flattened; no bristles or suckers.
      To 35mm long, usually 8 to 16mm.   White, brown, black etc.
      Moves by gliding over surfaces.   .... FLATWORMS - 9

5B   With a posterior sucking disc; mouth may be used as a sucker.
      With segments and annuli (finer rings); no bristles.
      Moves by looping using suckers;
        some can also swim.   ......   To p.39.   LEECHES - 15

5C   Typical worms; mostly thin and elongate, from 10mm to 300mm long.
      May have bristles along body.   Usually segmented.
      Moves by wriggling; some can swim; most burrow in mud etc.   - 6

   6A   Earthworm-like or thinner.   Reddish-brown.   10 to 100mm long.
         With segments and short bristles.   .... - 7

   6B   Small, without segments or bristles,
           or, long & thin (to 300x1mm) & with or without segments.- 8

   6C   Small; from 10 to 30mm long.   More or less transparent except
           for gut contents.
         With segments and bristles.
         Budding may give chains of individuals.   ....   NAIDID WORMS

7A   Earthworm-like in appearance
      and proportions.
      Posterior end square in section.   To 80mm.   *Eiseniella tetraedra*

7B   Thin (1.0-1.5mm diam.) and elongate (50-100mm long).   Active.
      Reddish-brown with greenish sheen.
      In mud under vegetation.   *Lumbriculus variegatus*

7C   As for B but smaller, usually under 40mm long.
      Curls up tight when disturbed. —
      Often present in high densities.   ..... TUBIFICID WORMS

   8A   Over 100mm long and about 1mm thick.   Usually almost white.
         No segments or bristles.
         Slow moving.   HAIRWORM - *Gordius*

   8B   Up to 300mm long, and thread-like.
         Nearly 500 segments and some bristles.   *Haplotaxis gordioides*

   8C   Smaller, some minute.   Both ends pointed.   Almost transparent.
         No segments or bristles.   — NEMATODE WORMS

CHECKLIST	NOTES

Dendrocoelum lacteum
*Bdellocephala punctata**

Phagocata vitta

Crenobia alpina

*Procerodes littoralis**

Dugesia polychroa

D.   *lugubris*

Planaria torva

Polycelis felina

P.   *nigra**
P.   *tenuis**

Bdellocephala punctata is similar in shape to *Dendrocoelum lacteum*
but is dark and mottled. It is large, reaching 35x8mm.   Recently
found in Ireland by T.K. McCarthy (*I.N.J.*, 1973, 17(12):419).

Procerodes littoralis has two eyes, rounded ears and oval body
with a neck.   Common, but only in brackish waters.

Polycelis nigra and *P. tenuis* can only be separated by microscopic
examination.   See F.B.A. Key (below), Fig. 5.

REFERENCES
   REYNOLDSON, T.B. 1967. *A Key to the British species of Freshwater
      Triclads*.   Freshwater Biol. Assn.   Sci. Publ. 23. 28pp.
   REYNOLDSON, T.B. and L.S. BELLAMY. 1970.   The status of *Dugesia
      lugubris* & *D. polychroa* (Turbellaria; Tricladida) in Britain.
      J. Zool., London, 162:157-177.
   SOUTHERN, R. 1936.   Turbellaria of Ireland.   *Proc. Roy. Irish
      Acad.*   43 B(5):43-72.

9A   One pair of eyes. .... -10

9B   Eyes numerous, around head end.   To 12mm long. -14

   10A   Body colour white; or, if over 20mm long then may be dark. - 11

   10B   Brown or tan, more rarely mottled or black. To 18mm long. - 12

11A   Large and broad; to 25mm long.
Eyes near anterior edge.
On stones etc., in lakes and canals.

Dendrocoelum lacteum

11B   Small and narrow; to 12mm long.
Eyes further back and close together.
In springs, also stony streams and some lakes.   *Phagocata vitta*

   12A   With long thin 'ears'.   Black.
Body elongate when moving.
In springs and pure cold streams.

Crenobia alpina

   12B   Without 'ears'.   'Eyes' large.   Broad.
Usually in standing waters. -13

13A   Head rounded; neck present; body solid.
Eyes close to anterior margin.
Ventral surface paler than dorsal surface.   *Dugesia polychroa*

13B   Head bluntly triangular; body slimmer than for A or C.
Eyes more posteriorly placed.
Colour similar above and below.   *Dugesia lugubris*

13C   Head almost square; no neck; body solid.
Eyes posteriad, as in B.
Ventral surface paler than dorsal.   *Planaria torva*

   14A   'Ears' prominent.   Usually brownish.
In springs and streams.   *Polycelis felina*

   14B   No 'ears'.   Usually black, but often
some specimens are brownish.
In lakes, ponds, rivers etc.   *Polycelis nigra* or *P. tenuis*

NOTES

Checklist – see page 40.

REFERENCE

MANN, K.H. 1964. *A Key to the British Freshwater Leeches with notes on their Ecology.* Freshwater Biol. Assn. Sci. Publ. 14. 50pp.

15A Body broad and flat (except *Piscicola*); head end narrow; and
        posterior sucker large.
      One, two or three pairs of eyes.
      Small; usually under 30mm long.
      Light coloured, or may have a dark mottled pattern. .....   - 16

15B Worm-like, or thick and flabby.
      Sides almost parallel.
      Eight or more eyes.
      Large; exceeds 40mm long.
      Brown or dark in colour; with or without colour pattern. ... - 20

   16A Head broader than 'neck'.
        Two pairs of eyes.
        Posterior sucker very large. .... - 17

   16B No 'neck' behind head.   Does not swim.
        One, or three pairs of eyes; rarely two.   SNAIL LEECHES - 18

17A Elongate body; rounded in cross-section.
      Second pair of eyes far apart.
      Active; can swim.

Piscicola geometra

17B Body broad and flattened.
      Eyes of about equal size.
      Does not swim.

Hemiclepsis marginata

   18A Small, usually about 10mm long; highly mobile.
        One pair of eyes and a 'scale' behind the head.
        Gray, with no colour pattern.
        Often found with young attached.   *scale*
        Active.   ....   *Helobdella stagnalis*

   18B Small; about 10mm long.
        Three pairs of eyes; anterior
           pair small and close-set.
        Clear amber colour; rarely with any markings.
        Inactive or slow moving.

Glossiphonia heteroclita

   18C Larger.   Three pairs of eyes (sometimes fewer).
        Brownish or greenish with dark markings.   .... - 19

CHECKLIST   NOTES

Piscicola geometra *

Hemiclepsis marginata *

Helobdella stagnalis *

Glossiphonia heteroclita *

G.   complanata

Batracobdella paludosa [++]

Boreobdella verrucata [++]

Theromyzon tessulatum

Haemopis sanguisuga

Hirudo medicinalis

*From p.39.

19A Colour pattern irregular and often clearly defined.
     Firm texture to the touch.
     Lake shores and stony streams,
          & may be with 19B or C...... ..... *Glossiphonia complanata*

19B Colour pattern less pronounced.
     Softer texture.
     More agile and extends further
          when moving. Quieter waters than A. *Batracobdella paludosa*

19C Colour pattern regular, with two large dark patches.
     Pigment lacking in head region.
     Rare species known from lake shores.

Boreobdella verrucata

20A Four pairs of eyes.
     Jelly-like texture.
     Large; to 40mm; body broad, head narrow.
     Dark, without pattern, or gut contents & internal organs show.
     Young appear transversely striped. *Theromyzon tessulatum*

20B Ten eyes, around the head.
     Thick and flabby.   Can swim.
     Large, to 100mm long; uniformly broad.
     May have yellow, orange or red markings. .... - 21

20C Eight eyes, in two rows.
     Earthworm-like.   Can swim.
     Slightly flattened & broader posteriorly.
     Colour reddish or dark brown. .... - 22

21A Colour dull; partly or completely mottled by dark markings.
     May have yellow-orange lateral stripes.
     Often found under stones out of water.

HORSE LEECH - *Haemopis sanguisuga*

21B Dark ground colour, with a red pattern
          in stripes along the back.
     Never has irregular mottled pattern.
     Has jaws which can pierce human skin.
     Highly contractile. ......   MEDICAL LEECH - *Hirudo medicinalis*

LEECHES Contd.

CHECKLIST   NOTES

Erpobdella octoculata

E.   testacea

Trocheta bykowskii

T.   subviridis[++]

Dina lineata

22A   A fine dark pattern present as irregular mottling dorsally,or, if
          generally dark, then with light marks at regular intervals.
       Under stones etc. in muddy or somewhat polluted waters.
       To 60mm long. *Erpobdella octoculata*

22B   Colour uniform; either light or dark brown, but may have feint
          stripes along its length.   ....   - 23

   23A   Longer than 60mm when extended naturally.
          May be found out of water.   .... *Trocheta* - 24

   23B   Shorter than above, usually under 50mm. ....   - 25

24A   Reddish-brown;
          largest specimens darkest.
       'Shine' on dorsal surface
          shows up numerous papillae (raised spots). *Trocheta bykowskii*

24B   Grayish-green or reddish with dull green sheen.
       Two feint dark lines may
          show along back; no
          papillae.   May be found
          far from open water. .... *Trocheta subviridis*

   25A   Reddish-brown, and may have feint dark lines along the back.
          Common in muddy habitats.
          To about 40mm long. *Dina lineata*

   25B   Uniformly brownish.
          Rare; found in temporary,
             weedy ponds.
          To about 50mm long.   .... *Erpobdella testacea*

MUSSELS AND SNAILS

Water snails have a lot of advantages to offer the student who wishes to really get to know how animals live and relate to their surroundings.

They are easily found for most of the year. Many species live only in shallow water since they must come to the surface from time to time to renew their air supply and all are slow moving and so cannot evade capture. Many will not leave the water and those that do will not move far if conditions are dry, so they are unlikely to escape from your custody. There are not so many species in Ireland that identifications get confused but, with practice, most can be distinguished on sight. All stages of their breeding cycle can be observed very readily for the egg-laying species. So why not have an aquarium of water snails? They are not only easy to feed, but can keep the aquarium clear of algae by eating it off the glass. Otherwise their shells are easy to collect and display. The mussels have many of the same advantages and filter floating algae from the water.

Since they cannot move fast, observe the snails in the water before disturbing them when you are out collecting. Try and count them, or see which surface they prefer - mud, stones, pond-weeds, reeds etc. Have you ever seen a sick or diseased one, or one being eaten? These questions may recur when you are looking at them in your aquarium and you may wish to remember the natural conditions so as to make your aquarium as suitable as possible for them.

Preservation may be wet or dry - the whole animal or just the shell. The animal has a shell so that it can recoil from unfavourable conditions therefore special killing techniques are needed to cause the animal to die "relaxed". One reason for a snail to expand to the full is to obtain oxygen when this is in short supply; and if it can't get enough it will die. So if you de-oxygenate some water and keep the snails in it and away from the air, they will die expanded and suitable for preservation in alcohol. Note that formalin can become acid and corrode the shells.

Usually there is no good reason for preserving the "flesh" so kill specimens you wish to preserve in water which is almost boiling and then remove the animal from the shell. Clean the shells if necessary and dry them. Keep them in labelled (match) boxes or tubes or fix them onto cards with water soluble glue - water soluble so that later they can be removed for any reason without damage. Series can be made to demonstrate growth - for example mount a collection as a histogram with numbers against each size category or add an average sized specimen every few weeks. In some species, notably the wandering snail (*Lymnaea peregra*), the form of the shell varies greatly from locality to locality so shells could be displayed to demonstrate this.   In all cases keep labels of locality and date with each specimen.

The large mussels lie almost buried in mud or gravel, the swan mussel in lakes and ponds etc., and the pearl mussel in fast flowing rivers.   The latter was once the basis of a lucrative pearl trade.   The tiny pea mussels may burrow in mud but the somewhat larger orb shells belong to active animals that can climb amongst vegetation like snails.

The large mussels and most operculate snails are either male or female but all other mussels and snails are both (hermaphrodite). Large numbers of eggs are produced by the female swan mussel; the pearl mussel has two or three times as many.   In fact the latter has about one million in each brood and they can live for ten to fifteen years.   This compares with a few young, five or so, in each brood in the pea mussel.   Five or a million!

Can we explain or understand this difference?   The net result of reproduction in a stable population is replacement - on balance each pea mussel or each female pearl mussel in its lifetime must leave one live young one which will in turn have young. If on balance they leave two then the population would double and if this kept on the population would run out of food or become otherwise out of balance with its environment.

A large part of the answer is that the pea mussels give birth to fully formed young, so few are needed - they have a high survival rate.   The pearl mussels release larvae which must attach to the gills of a fish in order to develop into young mussels.   Their chances of doing this are slim, hence their large numbers.   Similarly the larvae of swan mussels must attach themselves to the skin of a fish, usually on the fins and may be seen there as tiny black specks.

Except for one, all the snails lay eggs - several in a clear jelly - so their development can be observed using a microscope. The exception is Jenkin's spire shell which has live young; all are females - no males have ever been found.

Mussels have gills for obtaining oxygen from water directly. This is also the method used by the operculate snails which are of marine ancestry.   The other snails, from land ancestors, have a cavity under their shells in which they can keep air for breathing.

REFERENCES

JANUS, H. 1965. *The Young Specialist looks at Land & Freshwater Molluscs*. Burke. London. 180pp.

MACAN, T.T. 1969. *A Key to the British Fresh- and Brackish-water Gastropods*. Freshwater Biol. Assn. Sci. Publ. 13. 46pp.

CHECKLIST   NOTES

Margaritifera margaritifera

Anodonta cygnea
A.   anatina

Sphaerium corneum
S.   lacustre

Pisidium amnicum
P.   conventum
P.   moitessierianum
P.   casertanum
P.   henslowanum
P.   hibernicum
P.   lilljeborgi
P.   milium
P.   nitidum
P.   obtusale
P.   personatum
P.   pseudosphaerium
P.   pulchellum
P.   subtruncatum

REFERENCES
    ELLIS, A.E. 1946. Freshwater Bivalves (Mollusca) (*Corbicula,*
    Sphaerium, Dreissena). *Linnean Society of London Synopsis*
    of the British Fauna. 4. 15pp.
    ELLIS, A.E. 1947. Freshwater Bivalves (Mollusca)(*Unionacea*).
    Linnean Soc. of London Synopsis of the British Fauna. 5. 39pp.

MUSSELS AND SNAILS

1A   One or two hard shells; body soft.
Never with jointed legs or bristles. .... - 2

1B   Not as above. Try another key, p.3. -

2A   With two similar shells hinged together. .... MUSSELS - 3

2B   With one shell. .... FRESHWATER SNAILS - 6

3A   Large; 20 - 200mm across. .... PEARL or SWAN MUSSELS - 4

3B   Small; under 15mm across. .... PEA or ORB MUSSELS - 5

4A   Dark. Has teeth in the hinge between the shells.
To 120mm across. Elongate.
In fast flowing rivers.

hinge teeth

PEARL MUSSEL - *Margaritifera margaritifera*

4B   Usually light coloured. No hinge teeth.
To 200mm across. (In some localities all sps. are dwarfed.)
In slow rivers, canals,
lakes and ponds.

SWAN MUSSELS - *Anodonta*

5A   Outline of shell circular with umbo central.
Two siphons.
To 15mm across. ....

ORB SHELLS - *Sphaerium*

umbo

5B   Assymetrical or oval;
umbo usually not central.
One siphon only.
Adults 2 to 8mm across.

PEA SHELLS - *Pisidium*

CHECKLIST   NOTES

LIMPETS
Ancylus fluviatilis

Acroloxus lacustris

BLADDER SNAILS
Physa fontinalis
P.   *acuta*[+]
P.   *heterostropha*[*]

Aplexa hypnorum

*Not definitely known for Ireland.

6A   Not coiled.

..... LIMPETS - 7

6B   Coiled; either flat
       or as a spiral.

.... SNAILS - 8

7A   Almost circular; to 10mm diameter.
       Dark shell colour.
       In many habitats including lakes.
                RIVER LIMPET - *Ancylus fluviatilis*

7B   Elongate and flatter; to 6mm long.
       Light brown colour and thin shell.
       Lake shores and ponds.
                LAKE LIMPET - *Acroloxus lacustris*

8A   Coiled with a spire, that is, from a point.
       With or without an operculum.
                                            - 9
                                    operculum
8B   Coiled flat, that is, in one plane with no spire.
       With an operculum.   .... To p.53. - 16

8C   Coiled flat but without an operculum.   PLANORBID SNAILS - 11

9A   Coiled anticlockwise - therefore
          opening on left side.
       Shell thin and delicate.   No operculum.   BLADDER SNAILS - 10

9B   Coiled clockwise.   With an operculum.   OPERCULATE SNAILS - 15
                                              To p.53.
9C   Coiled clockwise.
       Without an operculum.   To p.55. - POND SNAILS - 19

10A   Spire short.   To 12mm long.
       In clean waters - lakes,
          canals and streams.
                                        BLADDER SNAIL -
                                        Physa fontinalis

10B   Spire elongate and slender.
       To 15mm long.
       In marsh and swamp conditions.
                MOSS BLADDER SNAIL - *Aplexa hypnorum*

CHECKLIST　　　　　　　　　　　　NOTES

Planorbarius corneus

Planorbis carinatus

P.　　　　*planorbis*

P.　　　　*albus*

P.　　　　*laevis*

P.　　　　*crista*

P.　　　　*leucostoma*

P.　　　　*vortex*

P.　　　　*contortus*

Segmentina complanata
S.　　　　*nitida*

11A   Whorls (coils) broad on both sides.
         Four (or five) whorls only.

                                    Right   &   *Left sides*

                                                 .... - 12

11B   Whorls narrow on right side, but
            may be narrow or broad on other side.
     Six or more whorls show on right side.

                                              5 3 1   - 14

     12A   Large and solid;
                 to 25mm diameter
                 and 12mm thick.

                                         RAMSHORN SNAIL -
                                  .... *Planorbarius corneus*

     12B   Less than 20mm diameter and thin.   .... - 13

13A   Medium sized; to about 15mm diameter.
     Shell with a keel (ridge) and
            slightly angular opening.
     Keel central or to one side as shown.

                     Planorbis carinatus or *P. planorbis*

13B   Small; to about 7mm diameter.
     About four whorls.   Thin.
     No keel; shell opening almost circular.

                       Planorbis albus or *P. laevis*

13C   Small, to about 5mm diameter.
     Outer whorl grows over inner whorls.
     No keel; shell opening V-shaped. ......

                                     Segmentina

13D   Tiny; to about 3mm diameter.
     Ridges cross the shell at short intervals.
     No keel; shell opening circular. ....

                           Planorbis crista

     14A   Whorls narrow on both sides.
           Mouth of shell almost circular or angled.
           Thin.   To 9mm diameter.

                 Planorbis leucostoma or *P. vortex*

     14B   Outer whorl broad on one side only.
           Mouth of shell flattened.
           Thicker.   To 6mm diameter. .... *Planorbis contortus*

CHECKLIST   NOTES

Theodoxus fluviatilis

Valvata cristata

V.   piscinalis

Bithynia tentaculata

B.   leachi

Potamopyrgus jenkinsi
*Hydrobia neglecta**

*H.   ulvae**

*H.   ventrosa**

*Pseudamnicola confusa**

*In brackish waters only.

REFERENCE
McMILLAN, N.F. & A.W. STELFOX. 1962. *Viviparus viviparus* (L.) in Ireland. *J. Conchology* 25(3): 117-121. [A possible addition.]

15A   Spire suppressed & shell limpet-like;
or else flattened, as in *Planorbis*.  ... - 16

15B   Spire with several whorls (coils). Shell at least 4mm broad. - 17

15C   Spire taller than opening of shell.   Under 3mm broad. - 18

16A   Shell limpet-like and solid -
more like a sea-shell.
Some specimens have a pattern.
To 12mm long.   THE NERITE - *Theodoxus fluviatilis*

16B   Shell flattened -
imperfectly coiled in one plane.
Delicate and small; to 4mm diameter. *Valvata cristata*

17A   Large; to 15mm tall.
Spire high with 5 to 6 whorls.
Mouth of shell oval/angled.   *Bithynia tentaculata*

17B   Smaller; to 7mm tall.
Spire relatively shorter; 4-5 whorls.
Mouth of shell almost circular.   *Bithynia leachi*

17C   Smaller again; to 5mm.   Spire blunt - its
breadth = height; about 4 whorls.
Mouth of shell circular.
Light in colour. .... *Valvata piscinalis*

18A   Whorls not greatly rounded.
Spire high; 5 or 6 whorls.
Black.   To 5mm tall.

JENKINS' SPIRE SHELL -
Potamopyrgus jenkinsi

18B-D   From brackish water or estuarine conditions
Note form of whorls.   Height 4-6mm.

B-   C-   D-

B = *Hydrobia ulvae*

C = *Hydrobia ventrosa*

D = *Pseudamnicola confusa*

POND SNAILS

CHECKLIST   NOTES

Succinea putris
S.   pfeifferi

Lymmaea stagnalis

L.   palustris

L.   truncatula

L.   glabra

L.   peregra

L.   auricularia

19A   Mouth of shell half or less of shell length.   *-spire*
Spire 1/4, or more, of shell length.

                      mouth-   .... - 20

19B   Mouth of shell longer than half of shell length.
Spire less than 1/5 of shell length.

                  body whorl   .... - 21

19C   Animal large for size of shell.
Shell almost transparent.
Mouth of shell symmetrically pear-shaped.
From aerial shoots of emergent vegetation
    and from marshes. .... - *Succinea*

    20A   Large; to 50mm; spire under half
        the length of the shell & pointed.
      Light horn colour.
      Large ponds, canals and lakes.
                 GREAT POND SNAIL - *Lymnaea stagnalis*

    20B   Medium size; to 20mm long;
        spire less pointed.
      Dark and may have streaky markings.
      Lake shores and ponds.   MARSH SNAIL - *Lymnaea palustris*

    20C   Small; to 10mm long;
        spire shorter; whorls more rounded.
      Dark colour.
      Usually found in damp conditions near
        ponds, rivers etc.   DWARF POND SNAIL - *Lymnaea truncatula*

    20D   Medium size; to 20mm long.
      Mouth of shell small - 1/3 of total length.
      Body whorl not much wider than
        preceding whorl.
      From small or temporary water bodies.   *Lymnaea glabra*

21A   To 25mm long - usually less.
Spire insignificant or moderately
    prominent; points backwards. Common.
Variable in most characters.   WANDERING SNAIL - *Lymnaea peregra*

21B   To 20mm long.   Mouth of shell very large.
Spire short and sharply pointed.
Spire points to the right
    of line of movement.   EAR POND SNAIL - *Lymnaea auricularia*

CRUSTACEANS, MITES AND SPIDERS

Water mites start life as larvae and grow attached to other animals, especially insects, but the adults are all free-living predators which swim or crawl and suck the body fluids from their prey. Most are brightly coloured or vividly patterned, often with red, yellow or green predominant. In size the adults range from minute to about 5mm diameter excluding the legs.

The raft spider ranges over the floating leaves of lakeside plants but *Argyroneta* is the true water spider. It lives, feeds, breeds and hibernates entirely underwater. It makes webs between the stems of aquatic plants and carries air down to fill each with a large bubble (or bell) to which it can retreat from hunting trips. It is common in some districts, especially in bog-pools, canals and ditches, and is easily kept in aquaria if these are set up specially to resemble the conditions in which water spiders are found.

The above species have come to live in freshwater from the land, but the crustaceans, below, have evolved from marine ancestors.

As elsewhere in these keys some brackish water species are included since most freshwater biologists sample estuarine reaches from time to time and find typical freshwater forms of life alongside some brackish water ones - for example, *Jäera, Sphaeroma* and *Corophium* among the crustaceans.

Ostracods, copepods and water-fleas are all small, rarely exceeding 2mm long, but they are common and often abundant. Many small or temporary pools have them because of their ability either to be dispersed or to withstand drought, frost or other harsh conditions. These characteristics are often due to resting eggs. Small crustaceans are a major source of food for many larger species of animals. They may be introduced to your aquarium for this purpose.

Remarkably, the three species of fish-lice have a lot in common. In fact they are related and have larvae (nauplii) like the small free-living copepods. All suck blood and are not particularly harmful to their hosts.

The salmon gill-louse infects adult salmon in freshwater, breeds, and more fish are infested before the kelts return to the sea. The adult louse can survive at sea on the fish's gills until the salmon returns to spawn a second time. The salmon sea-louse is more marine in habit but is found, with full egg-sacs, on salmon which have just arrived in freshwater. The eggs soon fall off and the adult dies. The fish-louse, *Argulus*, lives entirely in freshwater and unlike the others, the adult can swim and will attack a variety of fish species, from sticklebacks to pike.

The remaining crustaceans are all larger, and are important

food for fish. *Mysis* lives in deep lakes and although it swims it is rarely seen in shallow water. The crayfish is common in most districts, but is secretive and inoffensive, coming out at night to feed on small organisms and any dead matter. They are slow growing and live for several years.

Asellus is a genus with three Irish species. Two are widespread and often abundant especially in muddy or weedy conditions. The other is blind, in fact it has no eyes and lacks pigment. All asellids feed on detritus which is usually plentiful, but in caves, where the blind species lives, these decaying particles are one of the few forms of food.

Caves, like turloughs, are a special feature of Connacht but their faunas have not been investigated in detail. However, we know of the existence of one other blind, white crustacean from these underground waters. It is *Niphargus*, a relative of the freshwater shrimp, *Gammarus*. A more frequently seen member of the ground-water fauna is *Phagocata vitta*, a small white flatworm (p.37).

The distribution patterns of these species, incomplete as they are, have interesting and possibly important features. *Asellus aquaticus* may be extending its range since it survives better in organically polluted waters than does *A. meridianus* which is the species found on islands, such as Galway's Aran Islands. Here too is *Orchestia bottae* a species which, with *Gammarus tigrinus*, has spread inland in the English Midlands during the last few decades. *G. tigrinus* and *G. pulex* have been introduced to Lough Neagh and the latter species at least, is spreading and is now in neighbouring catchments. *G. pulex* is the most common and widespread species in Great Britain, a position held here by *G. duebeni* which is largely absent there. Recent records of *G. lacustris* are increasing its known range in western lakes.

Complete distribution patterns (see p.120) of these species would give a base-line for recording changes in distributions, which once explained could be informative about small but significant environmental or climatic changes.

REFERENCE

HYNES, H.B.N., T.T. MACAN & W.D. WILLIAMS. 1960. *A Key to the British species of Crustacea: Malacostraca occurring in fresh water*. Freshwater Biol. Assn. Sci. Publ. 19. 36pp.

CHECKLIST   NOTES

SPIDERS
  Argyroneta aquatica

  Dolomedes fimbriatus

HYDRACARINA - Over 200 species.

REFERENCE
  HALBERT, J.N. 1911. Clare Island Survey. Acarinida: i –
    Hydracarina. *Proc. Roy. Irish Acad.* 31 (39): 1-44.

CRUSTACEANS, MITES AND SPIDERS

1A   Animal globular or in two sections.
      With 4 pairs of legs for walking or swimming.

                              WATER SPIDERS and MITES - 2

1B   More than 4 pairs of limbs.   Mostly bottom-living.
      May have long antennae and large or small claws.
      Never with a tail ending in a single point.   *e.g.-*
      Adults over 10mm long.
                              LARGER CRUSTACEA - 4

1C   Small; mostly under 3mm long. Tail may end in a point.
      Usually able to swim in mid-water.
      Some "feelers" and legs are branched.   To p.63. - 10

1D   Rounded in outline.   10-15mm long.
      May have suckers for attaching on to a fish.   To p.67.
                              PARASITIC CRUSTACEA - 14

1E   Not as above.   Try another key (page 3).

    2A   Head and remainder of body as one unit; rounded.
          Usually small; up to 8mm diameter.
          Legs long or short.   Many swim well.
          Often coloured or patterned.
                              WATER MITES - HYDRACARINA

    2B   "Head" and remainder of body in two sections.
          Legs attached to "head" section. .... WATER SPIDERS - 3

3A   Black or dark brown without any light pattern.
      To about 12mm long.
      Often found underwater.

                    WATER SPIDER - *Argyroneta aquatica*

3B   Black or brown with yellow/orange stripes
          along the length of the body.
      Larger; to about 20mm long.
      Walks on water or floats on leaves
          on water surface.
                    RAFT SPIDER - *Dolomedes fimbriatus*

59

CHECKLIST   NOTES

ISOPODS
  *Jäera nordmanni**

  *Sphaeroma hookeri**
  *S.   rugicauda**

  Asellus aquaticus

  A.   meridianus

  A.   cavaticus

AMPHIPODS
  Orchestia bottae [+]

  Niphargus kochianus

  *Gammarus tigrinus**

  G.   lacustris
  G.   pulex

  G.   duebeni

*From brackish waters only.

4A   Body flattened from above.   ....   - 5

4B   Body flattened from the sides.   ..FRESHWATER SHRIMPS 7

4C   Body not appreciably flattened.   May have eyes on stalks.   - 9

    5A   Body elongate; to 12mm long.
            Legs and antennae long.   WATER LICE - 6

    5B   Body more circular; to 5mm only.
            Legs and antennae short.   .....   *Jaera nordmanni*

    5C   Able to swim & roll into a ball.
            To 16mm long.
            [Note hind appendages in A - C.]   *Sphaeroma*

6A-B Colour pattern on head continuous, or not, in mid-line.

      A=   B=   C=   A = *Asellus aquaticus*

                                B = *Asellus meridianus*

6C   No eyes or dark pigment or pattern.
      Blind cave animal.   ....   C = *Asellus cavaticus*

    7A   First pair of antennae short.
            Can hop.   *Orchestia bottae*

    7B   First & 2nd. antennae about equal in length.
            Yellowish with dark stripes.
                          Gammarus tigrinus

    7C   First antennae longer than 2nd.
            Does not hop.   .... - 8

8A   Eyes absent;
      blind cave dweller.
    "Hands" broad.
                    hand   *Niphargus kochianus*

8B   Eyes circular or oval.   .....   *Gammarus lacustris* or *G. pulex*

8C   Eyes kidney-shaped, & twice as long as broad.   *Gammarus duebeni*

CHECKLIST   NOTES

Astacus pallipes

Mysis relicta
*Neomysis integer**

*Corophium curvispinum**
*C.   volutator**

OSTRACODS
Many species, *e.g. Cypris* spp.

*In or near brackish waters.

9A   Large; to 100mm or over.
Solidly built with powerful claws.
Does not swim.

FRESHWATER CRAYFISH - *Astacus pallipes*

9B   Small; to 20mm.
Slender and without large claws.
Swims well.
Eyes on stalks.

GHOST SHRIMP - *Mysis relicta*

9C   Small; to 20mm.
One pair of stout appendages, almost as long as remainder of body,
    is carried in front of head.
Does not swim.
Eyes not on stalks.

head

Corophium

10A   Segmented structure obvious.
Streamlined; often with a forked tail.
Single eye.
May have 1 or 2 egg-sacs attached.

COPEPODS - 11

10B   Head often distinct and beaked.
Not usually streamlined.
One pair of large branched limbs -
    used for swimming with "hopping" movement.

WATER-FLEAS - 13

10C   Body enclosed by two valves = "shells".
Usually oval or kidney-shaped.
Limbs emerge from open valves
    during swimming.

OSTRACODS

CHECKLIST   NOTES

HARPACTICOID COPEPODS*

About 14spp., *e.g. Canthocampus* spp.

CALANOID COPEPODS

Diaptomus gracilis
D.   laciniatus
D.   laticeps
D.   wierzejskii
Eurytemora velox

CYCLOPOID COPEPODS

Cyclops affinis
C.   fimbriatus
C.   albidus
C.   fuscus
C.   gigas
C.   viridis
C.   bicuspidatus
C.   bisetosus
C.   crassicaudis
C.   languidoides
C.   nanus
C.   agilis
C.   bicolor
C.   leuckarti
C.   macruroides
C.   strenuus
C.   varicans
C.   venustus

CLADOCERA

See p.66.

*For parasitic copepods see p.66.

REFERENCES

HARDING, J.P. & W.A. SMITH. 1960. *A Key to the British Freshwater Cyclopoid and Calanoid Copepods*. Freshwater Biol. Assn. Sci. Publ. 18. 54pp.

O'RIORDAN, C.E. 1971. The Freshwater Copepod Work of G.P. Farran together with some other notes. *Proc. Roy. Irish Acad.* 71 B (6): 85-96.

11A   Body does not taper appreciably from head to tail.

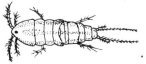

                                    .... HARPACTICOID COPEPODS

11B   Anterior unit (cephalothorax) much broader
        than tail segments.   .... - 12

    12A   Antennules ("feelers") with more than 17 segments.
            Eggs, if present, are carried in a single egg-sac.

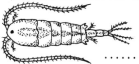

                          .... CALANOID COPEPODS

    12B   Antennules with fewer than 17 segments.
            Eggs, if present, in paired egg-sacs.

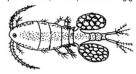

                                    .... CYCLOPOID COPEPODS

13A   Various shapes.   Rarely over 2mm long.
        "Tail" projection usually short.

                                    Typical CLADOCERA

13B-D   Atypical species; large and with long tails.
        May be extremely transparent.

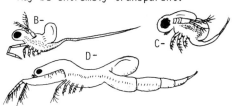

                                    B = *Bythotrephes*

                                    C = *Polyphemus pediculus*

                                    D = *Leptodora kindti*

CHECKLISTS   NOTES

PARASITIC CRUSTACEA
Argulus* coregoni
A.   foliaceus

Lepeophtheirus salmonis

Salmincola salmonea

CLADOCERA (From p.64.)
Bythotrephes longimanus
B.   cederstromi

Polyphemus pediculatus

Leptodora kindti

Typical cladocera – about 55 species.

*At least one species occurs in lakes & canals throughout Ireland.

REFERENCE
    SCOURFIELD, D.J. & J.P. HARDING. 1966. *A Key to the British Species of Freshwater Cladocera*. Freshwater Biol. Assn. Sci. Publ. 5. 55pp.

14A   Head and body circular; "tail" short.   To 7mm long.
Can swim using 4 pairs of branched limbs.
Two large suckers on under-side.

FISH-LOUSE - *Argulus*

14B   Head & body of two "regions"; egg-sacs may form two long "tails".
Can run over surface of fish using 4 pairs of unbranched limbs.
No suckers.   Found on salmon returning from the sea.
To 15mm without egg-sacs.

egg-sacs

SALMON SEA-LOUSE - *Lepeophtheirus salmonis*

14C   Present on gills of salmon.
To 10mm without
    egg-sacs.

egg-sacs

head

SALMON GILL-LOUSE - *Salmincola salmonea*

AQUATIC INSECTS

Insects have about as many species in freshwater as all the other groups combined and amongst the insects the caddis flies and the two-winged flies have most species. Consequently these are difficult to identify so they have barely been mentioned in the keys and checklists have not been given for them. However much of the work of freshwater biologists in Ireland has been done on the midges and fisheries biologists in particular know the various species of caddis in their larval stages. Bugs, dragonflies and mayflies have received some attention in Ireland but to date we rely to a considerable extent on the work of eminent British biologists for information, including distribution records, on many groups of freshwater animals.

Of all the insects that may be found living underwater only two groups are represented by adults. These are the bugs and beetles. The larvae of these aquatic adults are almost invariably aquatic themselves, but note that some adult bugs lack wings. Bugs and beetles are also very numerous in many terrestrial habitats in contrast to the mayflies, stoneflies, dragonflies and caddis of which every Irish species is totally aquatic in the larval stage. Yet all the adults of these groups are typical flying insects although a few will climb under water for egg-laying. A few species of moths have caterpillars that live in cases made of leaves, and some neuroptera (lace-wings) have fully aquatic young. The two-winged flies, unlike all the above groups except perhaps the bugs (where the young resemble the adults), show tremendous variety of form – larvae of mosquitoes with their breathing tubes, green or red tube-building chironomid midge larvae, hair-like biting midge larvae, the very transparent phantom midge larvae with buoyancy bladders for floating in mid-water, and various grub-like larvae including the rat-tailed maggot with its telescopic breathing tube. Not only do all these differ greatly in form from the adults but their pupae are different again and many are active and thus frequently caught. Mayflies, stoneflies and dragonflies do not have pupae since their larvae grow wings gradually in wing-buds and a final moult delivers the adults with body and legs not greatly changed and the wings simply take a few minutes to expand to full size and harden. In addition the mayflies have one more moult, from dun to spinner. The beetles and caddis have pupae but unlike those of the two-winged flies, they are usually hidden away and are not often found in casual collecting.

Many insects, especially those from canals or ponds, can live in small aquaria, or even dishes and jars, and not only can their movements be observed but also feeding, moulting and many other features of their lives. Also they are excellent subjects for simple experiments. Try removing caddis larvae from their cases (push, don't pull) and offer different specimens different materials for rebuilding – sand, broken shell, fine twigs or leaves etc. Time the intervals between visits to the surface for fresh air by beetles, using water at different (tolerable) temperatures or water with high and low concentrations of oxygen. Observe gill-movement in mayfly larvae under similar conditions. See how a wetting agent affects swimming in some active species or in bugs which live on or breathe at the water surface. Or, after observation of their natural habitat try and create conditions in an aquarium so as to keep a variety of species alive, and feeding for as long as possible.

A code for insect collecting* has recently been drawn up as a conservation measure to try and prevent the extinction of some of our most beautiful insects, especially butterflies and moths. Many of the 35 recommendations are for the experts who alone have the knowledge to apply them but the following extracts can be adhered to by all field workers for general collecting.

1.   No more specimens than are strictly required for any purpose should be killed.
2.   Readily identified insects should not be killed ... insects should be examined while alive and then released where they were captured.
3.   The same species should not be taken in numbers year after year from the same locality.
4.   When collecting on nature reserves, or sites of known interest to conservationists, supply a list of species to the appropriate authority.
5.   Do as little damage to the environment as possible. Remember the interests of other naturalists.
6.   Overturned stones and logs should be replaced in their original positions.

As yet we are not always certain about the actual rarity of some species which are infrequently collected, but even so it can best be assumed that the larger dragonflies and beetles are likely to be endangered by collecting. Therefore before deciding to take away, for example, a *Dytiscus* beetle think if you *need* to kill it or if you are prepared to return later to release it where it was found.

*A Code for Insect Collecting. 1973. Joint Committee for the Conservation of British Insects: 4pp. May be obtained from An Foras Forbartha, St. Martin's House, Waterloo Road, Dublin 4.

NOTES

AQUATIC INSECTS

1A With 3 pairs of legs and may have wings, wing-cases or wing-buds. Consists of head, thorax and abdomen; or maggot-like with about ten obvious segments; or intermediate. .... INSECTS - 2

1B Not an insect. Try another key, p.3.   -

2A Without wings or wing-cases, but may have wing-buds. Often with 1, 2 or 3 long and simple tail processes. May be maggot-like. .... INSECT LARVAE (NYMPHS) - 4

2B With wings; fore-wings modified as wing-cases with functional wings folded out of sight beneath them. Animal usually lives on water surface, or, is a good swimmer and lacks any tail processes (except *Nepa*, p.83). Front pair of legs may be specialised. Or, wingless but other wise of similar form. BUGS & BEETLES - 3

2C With one or two pairs of well-formed functional wings. Veins usually seen easily in wings in resting position; or moth-like. .... WINGED FLIES - To p.91. - 41

3A Wing-cases partly with veins showing & in part tougher; able to fold; and overlap across mid-line. Or, wingless but otherwise of similar form. ADULT & LARVAL BUGS - To p.83. - 24

3B Wing-cases uniformly tough; often shiny &/or with colour pattern; and meet in mid-line down the back. ADULT BEETLES - To p.85. - 27

4A Legs long - the last pair can extend back beyond posterior limits of the body. Tail processes absent; body stout & solid. - 7

4B Legs long, as above, but animal not as shown. Return to No. 3

4C Legs long, or short, (or absent) but not able to reach to, or beyond, posterior limits. Tail processes may be present & count as part of length. - 5

CHECKLIST   NOTES

BEETLES - See pp.84-88.

TRICHOPTERA = CADDIS
    Many species, in about 40 genera, are known for Ireland.

LEPIDOPTERA - Aquatic caterpillars.
    Acentropus niveus
    Cataclysta lemnata
    Nymphula nympheata
    N.   stagnata
    N.   stratiotata

REFERENCES
    HICKIN, N.F. 1967. *Caddis Larvae*. Hutchinson. London. 476pp.
    MOSLEY, M.E. 1939. *The British Caddis Flies (Trichoptera); a
        Collector's Handbook*. Routledge. London. 320pp. [Adults.]

5A   Legs well developed and obviously jointed. ....   - 6

5B   Legs absent or very short - e.g. maggots & midge larvae.   - 9

   6A   No long "tails". May have a pair of hooks posteriorly.
        (Long = one fifth of body length or more.) ....   - 7

   6B   One long tail process. ...... To p.77. - 14

   6C   Two long tail processes (not hooks). To p.77. - 15

   6D   Three long tail processes. To p.77. - 13

7A   Grub-like. May have a pair of hooks posteriorly.
     May live in a case made of leaves, sand, shells, etc.
     Legs of medium to short length.   Never with wing-buds.
                                          CADDIS LARVAE - 8

7B   Legs very long. Body stout and solid.   With a "mask".
     Wing buds usually present.
          To p.77.   STOUT-BODIED DRAGONFLY LARVAE - ANISOPTERA - 12

7C   Legs long or short.   May have tail processes of variable form.
     Never with wing-buds or posterior hooks.

                                    Various BEETLE LARVAE

   8A   Segments show clearly &/or
          posterior hooks large.
             WEB-SPINNING or FREE-LIVING CADDIS LARVAE

   8B   Found in a case.
        Abdomen soft &/or white.
                                    CASED CADDIS LARVAE

   8C   Legs very short. No posterior hooks.
        Case cut from a large leaf.
                   CHINA MARK MOTH CATERPILLAR - *Nymphula*

CHECKLIST   NOTES

DIPTERA - TWO-WINGED FLIES
Several hundred species may occur in Ireland. See p.90.

PLECOPTERA - STONEFLIES (From p.76.)
Large species (family PERLODIDAE)
   Perla bipunctata

   Dinocras cephalotes

   Perlodes microcephala

   Diura bicaudata

   Isoperla grammatica

Small species
   Amphinemura sulcicollis
   Brachyptera risi
   Capnia bifrons
   Chloroperla torrentium
   C.   *tripunctata*
   Leuctra fusca
   L.   *hippopus*
   L.   *inermis*
   Nemoura cinerea
   Nemurella picteti
   Protonemura meyeri
   P.   *praecox*
   Taeniopteryx nebulosa

NEUROPTERA Etc. + ALDER-FLIES Etc. (From p.76.)
   Sialis lutraria
   S.   *fulginosa**
   Osmylus fulvicephalus
   Sisyra dalii
   S.   *fuscata*
   S.   *terminalis*

*Not definitely known for Ireland.

REFERENCES - From p.76.
   HYNES, H.B.N. 1967. *A Key to the Adults and Nymphs of British Stoneflies (Plecoptera).* F-w. Biol. Assn. Sci. Publ.17. 91pp.

   KIMMINS, D.E. 1962. *Keys to the British Species of Aquatic Megaloptera and Neuroptera.* F.B.A. Sci. Publ. 8. 23pp.

9A   Head readily distinguishable or head end enlarged. .... - 10

9B   Head minute or otherwise not easily distinguished. .... - 11

10A-F   Examples of various types, as figured.

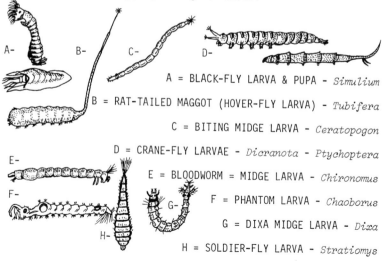

A = MOSQUITO (=GNAT) LARVA & PUPA - *Culex*

B = MOSQUITO LARVA & PUPA - *Anopheles*

C = BITING MIDGE PUPA - *Ceratopogon*

D = CRANE-FLY PUPA - *Ptychoptera*

E = NON-BITING MIDGE PUPA - *Chironomus*

F = PHANTOM MIDGE PUPA - *Chaoborus*

11A-H   Examples of various types, as figured below.

A = BLACK-FLY LARVA & PUPA - *Simulium*

B = RAT-TAILED MAGGOT (HOVER-FLY LARVA) - *Tubifera*

C = BITING MIDGE LARVA - *Ceratopogon*

D = CRANE-FLY LARVAE - *Dicranota* - *Ptychoptera*

E = BLOODWORM = MIDGE LARVA - *Chironomus*

F = PHANTOM LARVA - *Chaoborus*

G = DIXA MIDGE LARVA - *Dixa*

H = SOLDIER-FLY LARVA - *Stratiomys*

CHECKLIST

ODONATA - DRAGONFLIES
SHORT-BODIED ANISOPTERA
Cordulia arena
Libellula depressa
L.   fulva
L.   quadrimaculata
Orthetrum cancellatum
O.   coerulescens
Sympetrum fonscolombei
S.   nigrescens
S.   sanguineum
S.   scoticum
S.   striolatum
Somatochlora arctica

LONG-BODIED ANISOPTERA
Aeshna grandis
A.   juncea
Anax imperator
Brachytron pratense
Cordulegaster boltoni
Gomphus vulgatissimus

ZYGOPTERA
Agrion splendens
A.   virgo
Coenagrion puella
C.   pulchellum
Enallagma cyathigerum
Erythromma najas
Ischnura elegans
I.   pumilio
Lestes dryas
L.   sponsa
Pyrrhosoma nymphula

NOTES

STONEFLIES, ALDER-FLIES, Etc. - See p.74.

REFERENCES
FRAZER, F.C. 1949. *Handbook for the identification of British Insects: Odonata.* Royal Entomological Soc. of London. I(X): 48pp.
LONGFIELD, C. 1949. *The Dragonflies of the British Isles.* Warne. London.
MacNEILL, N. 1949. Distribution of Dragonflies in Ireland. *Irish Nat. J.* IX(9): 231-241.
MacNEILL, N. 1950. Dragonflies - A Key to the nymphs of Coenagriidae common in Ireland (Odonata - Zygoptera). *Irish Nat. J.* X(2): 32-38.

12A   Body broad.

A = Larva of SHORT-BODIED
       DRAGONFLY - ANISOPTERA

12B   Body stout but
         more elongate.

B = Larva of LONG-BODIED
       DRAGONFLY - ANISOPTERA

13A   Caudal processes leaf-like &
         easily broken off.
         Body slender.   Mask present - - - -
         No paired gills.   .... DAMSELFLY LARVA (NYMPH) - ZYGOPTERA

13B   Caudal processes thread-like & hairy.
         Paired gills, as plates or threads,
         may occur on abdomen.   .... MAYFLY LARVA (NYMPH) - 16

14A   Tail process fringed with hairs, as are
         seven pairs of abdominal appendages.
         Crawls.   Jaws of biting type.
         To 40mm long.   ....   ALDER-FLY LARVA - *Sialis*

14B   Similar appearance but hairs present
         in place of appendages.
         Broader and smaller; to 10mm long.
         Mouth parts piercing; lives on sponges.
         May be green.   ....   SPONGE-FLY LARVA - *Sisyra*

15A-E   Largest STONEFLIES (CREEPERS);to about 30mm(A-C) or 16mm(D&E).

A = *Perla bipunctata*

B = *Dinocras cephalotes*

C = *Perlodes microcephala*

D = *Diura bicaudata*

E = *Isoperla grammatica*

15F   Similar to above, with wing-buds developing.
         Smaller; to 12mm (body length).   Smaller species of STONEFLY

CHECKLIST*   NOTES

Ephemera danica

Caenis horaria
C.   moesta
C.   rivulorum

Ephemerella ignita
E.   notata

Ecdyonurus dispar
E.   torrentis
E.   venosus
E.   insignis

Rhithrogena haarupi
R.   semicolorata

Heptagenia fuscogrisea
H.   lateralis
H.   sulphurea

*This list is repeated for adult mayflies on pages 94 - 98.

REFERENCE

MACAN, T.T. 1970. *A Key to the Nymphs of British Species of Ephemeroptera with notes on their ecology.* Freshwater Biol. Assn. Sci. Publ. 20. 68pp.

16A   Gills double - ✱ & flexed over the back.
        Mouth parts show in front of head.
        Burrows in clean mud, sand or gravel.
        Can swim.   Whitish.
        Large; to 20mm long (body length less processes). *Ephemera danica*

16B   Not as in A.   ....   - 17

17A   Only one small pair of gills visible -
            others are covered by a flap.   *gill*
        Crawls.   In muddy habitats.
        Small; to 10mm long.   ....   *Caenis*

17B   More than two pairs of gills visible, either as plates or
            filaments etc.   ....   - 18

18A   Four pairs of gills visible.
        Each consists of an outer plate with
            filaments underneath.
        Gills are on the back and do not
            project over the sides.
        Crawls on vegetation etc.   Small; to 9mm long. .... *Ephemerella*

18B   With seven pairs of gills that project over the sides. ..... - 19

19A-C   Head and body flattened.
        Larger "joints" of limbs also flattened.
        Does not swim but clings to stones in flowing water.

        A-   B-   C-

        Compare pronotum
        of A - ⌒ and
        gills of C - ⬦
        with others.

        A = *Ecdyonurus*

        B = *Rhithrogena*

        C = *Heptagenia*

19D   Not particularly flattened.   Able to swim. .... - 20

CHECKLIST*   NOTES

Paraleptophlebia cincta
P.   submarginata

Leptophlebia marginata
L.   vespertina

Ameletus inopinatus

Siphlonurus armatus
S.   lacustris
S.   linneanus

Baëtis muticus
B.   niger
B.   rhodani
B.   scambus

Cloëon simile
Procloëon pseudorufulus
Centroptilum luteolum

*This list is repeated for adult mayflies on pages 94-98.

20A&B   "Tails" long.
        Swims awkwardly. A-   B-
        Gills double-

                                            A = *Paraleptophlebia*

                                            B = *Leptophlebia*

20C   "Tails" shorter than body.
        Swims well.
        Hairs absent on outer margins of lateral tail processes. ... - 21

    21A   Hind corners of last few abdominal
            segments drawn out into spines.
          All 3 "tails" of about the same length. .....   - 22

    21B   Last abdominal segments without spines.
          "Tails" of similar length or middle "tail"
            shorter than others.   .... - 23

22A   "Tail" processes held close together.
        All gills simple.

                                            Ameletus inopinatus

22B   "Tail" processes held more widely spread.
        Some gills with double plates.

            double   *single*   *Siphlonurus*

    23A   Middle "tail" process
            shorter than others.
          Gills all single and
            with rounded tips.

                                            Baëtis

    23B   All 3 "tails" of similar length.
          Gills may be all single & with pointed tips,
            or, some may be double.
          Segmentation of "tails" accentuated
            by dark colour pattern.   *Cloëon, Procloëon* or *Centroptilum*

CHECKLIST   NOTES

Nepa cinerea

Hydrometra stagnorum

Gerris argentatus
G.   lacustris
G.   lateralis
G.   najas
G.   odontagaster
G.   rufoscutellatus
G.   thoracicus

Velia* capria
V.   saulii

Microvelia pygmaea
M.   reticulata

Hebrus ruficeps

Mesovelia furcata

Notonecta glauca
N.   maculata
N.   obliqua
N.   viridis

Plea leachi

Ilyocoris cimicoides

Aphelocheirus aestivalis

CORIXIDAE - See p.84.

  *At least one species occurs throughout Ireland.

REFERENCES
    HALBERT, J.N.   1935.   A list of the Irish Hemiptera (Heteroptera
        and Cicadina).   *Proc. Roy. Irish Acad.* 42 B (8): 207-318.
    MacNEILL, N.   1973.   A revised and tabulated list of the Irish
        Hemiptera-Heteroptera - Part I. Geocorisae. *Proc. Roy. Irish
        Acad.* 73 B (3):   57-60.   [Introduction and notes - see
        Acknowledgements p.4.]
    MACAN, T.T.   1965.   *A revised Key to the British Water Bugs (Hemi-
        ptera-Heteroptera).* Freshwater Biol. Assn. Sci. Publ. 16. 78pp.

24A   Legs very long; animal adapted for movement on water surface.- 25

24B   Active swimmer under water.   Swimming hairs on hind legs.   - 26

24C   Crawls under water.
       Adult has a long
           breathing tube.
       Leaf-like.   To 30mm.   WATER-SCORPION
                                                      Nepa cinerea

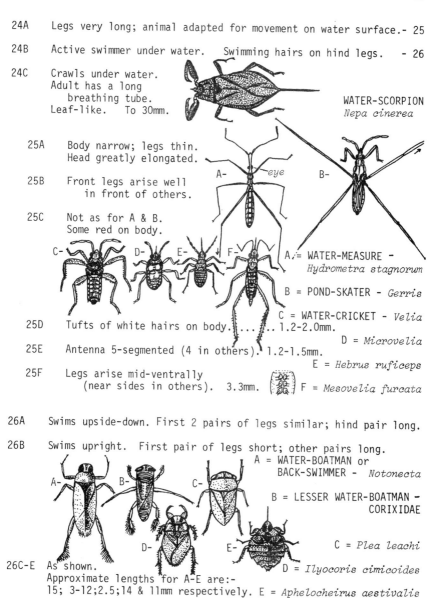

   25A   Body narrow; legs thin.
          Head greatly elongated.

   25B   Front legs arise well
             in front of others.

   25C   Not as for A & B.
          Some red on body.

                                            A = WATER-MEASURE -
                                               Hydrometra stagnorum

                                            B = POND-SKATER - *Gerris*

                                            C = WATER-CRICKET - *Velia*
   25D   Tufts of white hairs on body..... .. 1.2-2.0mm.
                                                      D = *Microvelia*
   25E   Antenna 5-segmented (4 in others). 1.2-1.5mm.
                                                    E = *Hebrus ruficeps*
   25F   Legs arise mid-ventrally
             (near sides in others). 3.3mm.   F = *Mesovelia furcata*

26A   Swims upside-down. First 2 pairs of legs similar; hind pair long.

26B   Swims upright. First pair of legs short; other pairs long.
                                         A = WATER-BOATMAN or
                                             BACK-SWIMMER - *Notonecta*

                                         B = LESSER WATER-BOATMAN -
                                                        CORIXIDAE

                                               C = *Plea leachi*

26C-E As shown.   D = *Ilyocoris cimicoides*
       Approximate lengths for A-E are:-
       15; 3-12;2.5;14 & 11mm respectively. E = *Aphelocheirus aestivalis*

CHECKLIST

Dytiscus semisulcatus

D.   *marginalis*
D.   *circumcinctus*
D.   *lapponicus*

Acilius sulcatus
A.   *canaliculatus*
Hydaticus seminiger

Colymbetes fuscus

(*Ilybius* – see p.86.)

CORIXIDAE (from p.82.)
  Cymatia bonsdorffi
  Glaenocorisa propinqua
  Corixa punctata
  C.   *dentipes*
  C.   *panzeri*
  C.   *affinis*
  Hesperocorixa linnei
  H.   *sahlbergi*
  H.   *castanea*
  H.   *moesta*
  Arctocorisa germari
  Callicorixa praeusta
  Micronecta minutissima
  M.   *poweri*
  Sigara – 13 species.

REFERENCES
    BALFOUR-BROWNE, F.   1940 & 1950 (2 volumes). *British Waterbeetles*.
       Ray Society. London.
    BALFOUR-BROWNE, F.   1953. *Handbook· for the Identification of
       British Insects. Coleoptera Hydradephaga*. Royal Entomological
       Soc. of London. IV (3):33pp.
    HOLLAND, D.G. 1972. *A Key to the Larvae, Pupae and Adults
       of the British species of Elminthidae*. Freshwater Biol.
       Assn. Sci. Publ. 26.
    LINSSEN, E.F. 1959. *Beetles of the British Isles*. First Series
       [vol. 1]. Warne. London. 300pp.

27A   Good swimmer; 5 - 35mm long. Body streamlined.
Hind legs the longest and fringed with hairs for swimming. - 28

27B   Good swimmer, but small; to 5mm long or, if longer (to 8mm), with
first pair of legs longest; otherwise last pair longest. - 34

27C   Does not swim; crawls on submerged surfaces.   To p.89. - 38

28A   Over 22mm long. Sides with broad yellow stripes. .... - 29

28B   Between 13 & 20mm long. Black &/or darkish brown. ... - 30

28C   Under 12mm long. .... - 31

29A   Undersurface black.
No yellow between the   A-
lateral stripes.
Males & females similar.
24 - 30mm long.

29B   Undersurface yellowish.
Area between head and wing-
covers ringed with yellow.
Males shiny; females ridged.
26 - 32mm long.

A = *Dytiscus semisulcatus*

B = GREAT DIVING BEETLE
Dytiscus marginalis

30A   Body much flattened.
Brownish.   A-
About 15mm long.
Females ridged.

30B   Less flattened.
Olive-brown, with black head.
About 16mm long.

A = *Acilius sulcatus*

B = *Colymbetes fuscus*

30C   Not flattened. Black, with small orange
dots on each wing-case. About 14mm long.

C = MUD-DWELLER
Ilybius ater

31A   Predominantly black, with or without some colour pattern. - 32

31B   Predominantly yellowish, reddish or brownish colour. ..... - 33

CHECKLIST   NOTES

*Ilybius ater**

I.   fuliginosus

Agabus bipustulatus

A.   unguicularis

A.   nebulosus

A.   conspersus

A.   sturmii
Also:-
   Ilybius aenescens
   I.   obscurus
   Agabus - 8 spp.
   Rantus - 5 spp.
   & *Copelatus agilis*

Platambus maculatus

Hydroporus planus

H.   pubescens

H.   palustris
Also:-
   Hydroporus - over 20 spp.
   Deronectes assimilis
   D.   depressus
   D.   duodecimpustulatus
   D.   griseostriatus
   Laccornis oblongus
   Oreodytes borealis
   O.   rivalis
   O.   septentrionalis
   Noterus clavicornis
   N.   `capricornis
   Laccophilus hyalinus
   L.   minutis
   & *Stictonotus lepidus*

PRONOTUM - the dorsal area of thorax between head & wing-cases.

*From p.84.

32A   Black all over except for 2 small red dots on the head; &
       antennae and some mouth parts are reddish.
       About 10mm long.

32B   Black, with outer borders of
       wing-cases orange/yellow.
       About 8mm long; more slender.

A = *Agabus bipustulatus*

32C   Shining black all over.
       Sides almost parallel & body   B = *Ilybius fuliginosus*
       domed in cross section.
       Smaller again at about 6mm long.   C = *Agabus unguicularis*

33A   Marbled pattern of yellow/gold and black; head black;
       pronotum (the back between head & wing-cases)light coloured
       with two black dots.
       About 8-8.5mm long.

33B   Similar to A but slightly smaller.

33C   Similar to A but smaller (7mm) &
       pronotum black with   A = *Agabus nebulosus*
       brown edges.
                                                B = *Agabus conspersus*
33D   Pattern more symmetrical.
       Pronotum light surrounded   C = *Agabus sturmii*
       by dark colour.
       About 8mm long.   D = *Platambus maculatus*

34A   Front legs longest; hind legs short & flattened.
       Black and shiny.   Often swims in surface film. 4-8mm long. - 37

34B   Similar to *Dytiscus* in shape.   From 2 to 4mm long.
       Pronotum black or darker than wing-cases.   .... - 35

34C   Body more spherical &/or legs relatively longer than for B.
       Pronotum usually lighter in colour than wing-cases.   2-5mm. - 36

35A   Broad; dark & dull all over.
       3.5mm long.   A =
                                                Hydroporus planus

35B   Shiny-black all over, or wing-
       cases red or yellow. 2.5mm.   B = *Hydroporus pubescens*

35C   Dark with 4 orange or yellow marks. 2.5mm. *Hydroporus palustris*

CHECKLIST   NOTES

Hygrotus inaequalis
H.   confluens
H.   impressopunctatus
H.   novemlineatus
H.   quinquelineatus
Bidessus minutissimus

Hyphydrus ovatus

Haliplus apicalis
H.   confinis
H.   flavicollis
H.   fluviatilis
H.   fulvus
H.   immaculatus
H.   lineatocollis
H.   obliquus
H.   ruficollis
H.   variegatus
H.   wehnckei
Hygrobia hermanni
Brychius elevatus
Peltodytes caesus

Orectochilus villosus

Gyrinus bicolor
G.   caspius
G.   colymbus
G.   marinus
G.   minutus
G.   natator
G.   urinator

Helophorus aquaticus
H.   flavipes
H.   brevipalpis
H.   minutis

Hydrobius fuscipes
H.   niger

Anacaena globulus
A.   limbata
Laccobius biguttatus
L.   minutus

36A   Distinct markings on wing-cases.
        Body rounded and deep.
        About 2.5mm long.   A-   B-

36B   Almost uniformly light
           tan coloured.
        Spherical also, but larger;
           about 4.5mm long.   C-   A = *Hygrotus inaequalis*

36C   Legs long; swims awkwardly.   B = *Hyphydrus ovatus*
        Head small.   Finely marked.
        2.7 to 4.3mm long.   C = *Haliplus*

    37A   Covered with fine hairs on upper surface.
            Narrow bodied.   HAIRY WHIRLIGIG -
            5 - 6mm long.   ....   *Orectochilus villosus*

    37B   Black & shiny without fine hairs.
            Varies from 3 to 8mm long.
            Often two or more species found
               swimming together. .... WHIRLIGIG BEETLE - *Gyrinus*

38A   Narrow bodied; not having the appearance of being adapted to an
           aquatic life.   ....   - 39

38B   Circular or sub-circular in outline; domed back, flat underneath.
        Legs sometimes difficult to unfold from under body.   ...... - 40

    39A   Yellowish brown.
            Wing-cases ridged.
            Hairy underneath.
            From 6 to 7mm long.   .... *Helophorus aquaticus*

    39B   Similar to A but smaller;
               about 3mm long. .... Possibly *Helophorus flavipes*

40A   Black and shiny with rows of pits
           on wing-cases.
        Legs thin, brown & partly hairy.
        From 6 to 8mm long.   .... *Hydrobius fuscipes*

40B   Smaller; (2-)3mm long.
        Head & pronotum black;
           wing-cases brown with lighter margins. .... *Anacaena globulus*

CHECKLIST   NOTES

DIPTERA[*]– TWO-WINGED FLIES   (Sample genera)
Anopheles

Culex
Aëdes
Theobaldia

Chaoborus

Dixa

Chironomus

Ceratopogon

Tipula
Dicranota
Ptychoptera

Stratiomys

Tubifera

Simulium

*Several hundred species may occur in Ireland.
See p.75 for aquatic stages.

REFERENCES

DAVIES, L. 1968. *A Key to the British Species of Simuliidae (Diptera) in the larval, pupal and adult stages.* Freshwater Biol. Assn. Sci. Publ. 24. 126pp.

FAHY, E. 1972. A preliminary account of the Simuliidae (Diptera) in Ireland, with observations on the growth of three species. *Proc. Roy. Irish Acad.* 72 B (6): 75-81.

GRIMSHAW, P.H. 1912. Clare Island Survey. Diptera. *Proc. Roy. Irish Acad.* 31 (25): 1-34.

MURRAY, D.A. 1972. A list of the Chironomidae (Diptera) known to occur in Ireland, with notes on their distribution. *Proc. Roy. Irish Acad.* 72 B (16): 275-293.

41A   With one pair of wings.   .... - 42
41B   With two pairs of wings.   Often large, or moth-like. ..... - 45

42A   Small (under 5mm) & stout; or longer (to 15mm) & slender. - 43
42B   About 10 - 15mm long & stout.
42C   Long bodied (over 15mm) and slender.

B-

C-

B = HOVERFLY
    Tubifera

C = CRANE-FLY
    Tipula

43A   Antennae short and fine. 2 or 3 "tail" processes. To p.95. - 50
43B   Antennae hairy. No "tail" processes.   .... - 44

44A   Small (about 3mm long) & of solid build.
44B-C Medium size (to 12mm).
       With long sucking mouth parts.

A-

B-   C-

D-

B-   C-

A = BLACK-FLY
    Simulium

B = CULICINE MOSQUITO
    Culex

C = ANOPHELINE MOSQUITO
    Anopheles

44D   Similar to mosquitoes but
       without long sucking mouth parts.

D = NON-BITING MIDGE
    Chironomus

CHECKLIST

DRAGONFLIES are listed on p.76.

CADDIS FLIES - see references p.72.
   Sedge
   Great Red Sedge
   Green Sedge
   Gray Flag
   Grannom

ALDER-FLIES Etc. - larvae are partially keyed out on p.77;
                 checklist is on p.74.

   Sponge-Fly
   Alder-Fly
   Lace-wing

   STONEFLIES - larvae are partially keyed out on p.77;
               checklist is on p.74.

   Yellow Sally
   Willow-fly
   Needle-fly
   February Red
   Early Brown

NOTES

The names given here are the *Anglers' Names* for the adult flies of some well known species.

REFERENCE
   HARRIS, J.R. 1952 (Revised Edition, 1956). *An Angler's Entomology*. Collins. London. 268pp.

45A   Fore- and hind-wings very similar in size and shape. ..... - 46

45B   Hind-wings less than half the length of the fore-wings. .... - 50

46A   Antennae short. Eyes large.   Wings long & do not fold.
                                     DRAGONFLIES - ODONATA - 47

46B   Antennae long.   Eyes not noticeably large.
       Wings fold over the back when animal is resting such that not
       all four wings can be seen. .... - 48

47A   Eyes almost meet across head.   Body heavy and solid.
       Usually over 50mm long.

47B   Eyes wide apart.
       Body slender.
       Usually under 50mm long.

       A-   B-

       A = DRAGONFLIES - ANISOPTERA; B = DAMSELFLIES - ZYGOPTERA

48A   No "tail" processes.   Wings, at rest, meet at an angle over
       the back and are longer than the fly's body.
       Wings hairy; usually with
       a feint pattern.
       Moth-like but without a "tongue".

                                   CADDIS FLIES - TRICHOPTERA

48B   Wings not hairy; veins easily visible.   .... - 49

49A   Wings much longer than the body &, when held at rest almost form
       a semicircle and are vertical on each side of the body, & have
       numerous veins.

49B   Wings long with numerous veins   A-
       but held forming a broad
       "roof" over the body.   B-
       No "tail" processes.   A = SPONGE-FLIES - *Sisyra*
                                                B = ALDER-FLIES - *Sialis*
49C   Wing area less; fewer veins.
       Wings sometimes shorter than body.
       Two "tail" processes. Similar to larva. STONEFLIES - PLECOPTERA

CHECKLIST*   NOTES

Caenis horaria
C.   *moesta*
C.   *rivulorum*

Siphlonurus armatus
S.   *lacustris*
S.   *linneanus*
Ameletus inopinatus

(*Heptagenia* – see p.96.)

*These names are repeated for mayfly larvae on pages 78 & 80.

REFERENCE
    KIMMINS, D.E. 1954. *A revised key to the adults of the British species of Ephemeroptera.* Freshwater Biol. Assn. Sci. Publ. 15. 71pp.

50A   One pair of wings only, or hind pair minute. 2 or 3 "tails". - 51

50B   Two pairs of wings.   Two or 3 "tail" processes.   .... - 52

51A   Three "tail" processes.
        Wings whitish.
        Fly small.   ANGLERS' CURSE - *Caenis*

51B   Two "tail" processes.   ....   To p.99. - 60

52A   Two "tail" processes.   .... - 53

52B   Three "tails".   .... To p.99. - 62

53A   Hind-wing one third or more of length of fore-wing.
        Medium or large size.   .... · 54

53B   Hind-wing relatively small; one sixth or less of fore-wing.
        Medium or small fly.   .... - 58

54A   Body slender; not flattened.
        Wings long & slender.
        Large.   SUMMER MAYFLY
        Found May to August near lakes, ponds etc.   *Siphlonurus*

54B   Body wide & somewhat flattened.
        Wings not markedly elongate.   .... - 55

55A   With a small dark band on each leg "joint".
                                                   ...... - 56

55B   With small reddish bands marking the "thigh joints".
        Body brownish.
        Found May & June near rocky limestone lakes.

                                         BROWN MAY DUN & SPINNER
                                         Heptagenia fuscogrisea

55C   No colour bands on leg "joints".   .... - 57

CHECKLIST *   NOTES

Rhithrogena semicolorata

R.   haarupi

Heptagenia sulphurea

H.   lateralis

*H.   fuscogrisea***

Ecdyonurus dispar
E.   torrentis
E.   venosus
E.   insignis

Baëtis muticus

B.   rhodani
B.   niger
B.   scambus

*These names are repeated for mayfly larvae on pages 78 & 80.
**From p.94.

56A   Wings plain; bluish-gray;
            hind-wings paler than fore-wings.
        Medium size.
        Found May to August near
            fast stony rivers & streams.   OLIVE UPRIGHT DUN &
                YELLOW UPRIGHT SPINNER - *Rhithrogena semicolorata*

56B   Wings with mottled pattern; brownish fawn colour.
        Larger than A.
        Found March & April (& May).   MARCH BROWN -
                                                Rhithrogena haarupi

    57A   Flies all yellow except for eyes   A-   B-
                which are black or electric blue.
            Found May to August near fast rivers
                and rocky limestone lakes.
                                                A = YELLOW MAY DUN & SPINNER -
    57B   Wings dark gray and plain.   *Heptagenia sulphurea*
            Body dark with a bright yellow
                streak in front of each wing root. B = DARK DUN &
            Found May to September on stony shores   THE SPINNER -
                of large lakes and near rapid streams. *Heptagenia lateralis*

    57C   Wings fawn colour.   THE DUN
            Body drab brown or dark red.   LATE MARCH BROWN
            Found March to October near   GREAT RED SPINNER
                stony streams and rivers.   *Ecdyonurus*

58A   Hind-wings oval.
        Fore-wings with paired marginal
            veins between main veins. .... - 59

58B   Hind-wings very narrow or absent.
        Fore-wings with single marginal veins
            between main veins.   .... - 60

    59A   Small flies.
            Wings gray-black or blue-black (but male spinners colourless).
            Body dark brown or bluish-black.   IRON-BLUE DUN & SPINNER
            "Tails" white or gray.   LITTLE CLARET SPINNER
            Found April to September near flowing water.   *Baëtis muticus*

    59B   Medium-sized fly. Lighter colours than for A.
            Reddish-brown rings on "tails". LARGE DARK OLIVE DUN & SPINNER
            Spring & late autumn etc.   (& SMALL & MEDIUM OLIVES) -
            Fast flowing waters. ....   *Baëtis rhodani*

CHECKLIST*   NOTES

Centroptilum luteolum

Procloëon pseudorufulum

Cloëon simile

Ephemera danica

Ephemerella notata

E.   ignita

Leptophlebia vespertina
L.   marginata
Paraleptophlebia cincta
P.   submarginata

*These names are repeated for mayfly larvae on pages 78 & 80.

60A   Hind-wings minute.   Fly small.
Fore-wings pale or medium blue-gray.
Eyes orange, or brownish-red.   LITTLE SKY-BLUE DUN & SPINNER
Found April to November; widely   PALE WATERY DUN & SPINNER
   distributed on lake shores & streams.   *Centroptilum luteolum*

60B   Hind-wings absent; fore-wings light/dark gray or colourless. - 61

61A   Wings grayish-white; may have bright green tint at their base.
Body honey coloured or pale yellow.
With 6 - 8 cross-veins in stigmatic area -
Found May to October near slow rivers.
      PALE EVENING DUN & SPINNER - *Procloëon pseudorufulum*

61B   Wings medium or dark mousey-gray & often tinged with green or
   pale olive along main veins or, in female, colourless with
   brownish-olive body.
With 9 - 11 cross-veins in stigmatic area -
Found March to November near slow rivers & lakes.
      LAKE OLIVE DUN & SPINNER - *Cloëon simile*

62A   Large fly.
Wings greenish-gray with dark markings.
Body whitish or cream with dark brown marks.
Found April to September on lakes & flowing water.   THE MAYFLY
   GREENDRAKE (DUN) & SPENT GNAT (SPINNER)   *Ephemera danica*

62B   Medium sized fly. .... - 63

63A   Flies all yellow or partly orange behind the head.
Found May and June near moderately fast rivers.
      YELLOW EVENING DUN & SPINNER - *Ephemerella notata*

63B   Olive-green, brownish-orange
   or pale orange coloured body.
Wings darkish blue-gray.   BLUE-WINGED OLIVE DUN & SPINNER
From April to September by   BLUE-WINGED SHERRY SPINNER
   fast streams & rivers.   *Ephemerella ignita*

63C   Blackish-brown or dark
   claret body colour.
Very dark blue-grey fore-wings;
   hind-wings much paler.   CLARET DUN &
Found April to August on lakes,   LARGE CLARET SPINNER -
   tarns & small stony streams.   *Leptophlebia vespertina*

FISHES AND AMPHIBIANS

Some of our best known fishes live in freshwater for part of their lives and in the sea at other times. Their migrations therefore require them to pass through estuaries twice in each generation. Here they are conviently caught or are otherwise easily affected by man's activities. Some of our least known species, such as the shads, are also migratory. In fact the lives of about a dozen of our river fish species depend on access from the sea.

In spring the large sea lampreys may be seen building a nest in the shallower stony areas of rivers just above the estuarine reaches. They will spawn and die and the young ammocoete larvae will spend several years in freshwater feeding on fine food particles from the mud. Eventually they will change to their adult form and go to sea where they feed on fish. The brook lamprey has a similar life-history but does not feed on fish and can be found inland because it does not go to sea at any stage.

Young eels are about two years old when they reach our shores, in winter and spring. They enter rivers as glass eels and spend some time in the lower reaches. These eelogs can climb wet vertical surfaces and do so if necessary to get into inland waters. After feeding for six to ten years as yellow eels they become silver eels, and during autumn migrate to sea to swim to their spawning ground off the south-east of North America. Despite the several stages, we have but one species.

Sturgeons were never common but records have become less frequent than formerly. They again live and grow at sea and, as enormous fish, require to enter a river to spawn. Ireland is on the periphery of the range of these Asian fish and they may never breed here.

The shads are herring-like fishes but with life histories more like sea-trout for they grow in estuarine and inshore waters but in spring ascend the larger rivers to spawn. The allis shad can be large, to 8lbs. weight, but the size is dependant on conditions. A population of the twaite shad is landlocked in the Killarney Lakes and averages less than half the usual size for sea-run fish of the same species. Shads were once caught for food, at least in the Shannon, but are now rare. The smaller smelt is an estuarine salmonid that lives in shoals but is known only for the Shannon.

The tiny estuarine goby can be found in the mouths of many rivers and here too shoals of mullet of different age-groups come and go as does the flounder. Our two species of sticklebacks can live in brackish waters but most live entirely in freshwater. The ten-spined stickleback is infrequent but the three-spined sprick or tiddler is common and widespread in weedy waters.

All our larger rivers are important for the salmon that ascend them from the sea.   Many of these large fish are caught but the remainder spawn the following winter in the rivers and the young parr grow in freshwater.   After about two years, usually during May, they migrate to the sea as silvery smolts about 25cm long.   Most return a year later weighing 5 to 6lbs; these are termed grilse.   Others, a declining proportion, spend longer at sea and return as even bigger fish, usually earlier in the year. Many grilse and salmon get back to the sea after spawning (kelts) and after a variable time some reappear in the rivers to spawn again.   The grilse and larger salmon do not feed in freshwater. (See p.67 for some parasites of salmon.)

Much concern is being expressed about the future of our salmon runs and this is paralleled by increased fishing and increasing pollution of various types of our rivers and estuaries.   Our endeavours to conserve salmon should be linked to efforts to conserve a worthwhile human environment.   If we fail with salmon then the latter, and ultimate, goal will have become harder to attain.

Apart from a few small lakes where the American rainbow trout has been stocked, all our trout, in lakes, rivers and estuaries, are of one native species - the brown trout.   Two other salmonid fishes are confined to freshwater.   The pollan or whitefish is found only in Lough Neagh, Lough Erne and some Shannon lakes and the alpine char occurs in a larger number of lakes.

Salmon, pollan, alpine char and trout adapt noticeably in many ways to their environment.   Depending on habitat conditions their size, shape, colouring and habits vary within broad limits. The brown trout is the most plastic, hence the changes that are rung on the fads and fashions of anglers as to the relative merits of different stocks and on appropriate management policies.   This adds spice to the sport of angling but greatly complicates conservation efforts.

All the remaining fish live entirely in freshwater.   Thus they did not reach Ireland themselves by sea.   Some will have arrived when Ireland was connected by land to Europe, but others, including the pike, carp, goldfish, roach and dace have been introduced comparatively recently and for various reasons.

Many of the smaller fishes are very suitable for keeping in aquaria - the stone loach, gudgeon, minnow, goldfish, and of course the ever-popular stickleback.   The young of other species can also live in aquaria; small eels and trout can be kept for years.

Our two common amphibians, the frog and the newt, are often caught in spring at their breeding places.   Given big enough aquaria and conditions that approach their habitat requirements they too can live for long periods, both as tadpoles and adults.

REFERENCES - See p.102.

CHECKLIST　　　　　　　　　　　　　　NOTES

Petromyzon marinus

P.　　　　fluviatilis

Lampetra planeri

REFERENCES - From p.101.
　　BAGENAL, T.B.　1971.　*The Observer's Book of Freshwater Fishes.*
　　　　Warne.　London.
　　MAITLAND, P.S.　1972.　*A Key to the British Freshwater Fishes.*
　　　　Freshwater Biological Association.
　　O'RIORDAN, C.E.　1965.　*A Catalogue of the Collection of Irish
　　　　Fishes in the National Museum of Ireland.*　Stationery Office.
　　　　Dublin.　96pp.
　　WENT, A.E.J. & M. KENNEDY.　1969.　*List of Irish Fishes.*
　　　　Stationery Office.　Dublin.　44pp.

FISHES AND AMPHIBIANS

1A   Small; about 10mm long. May be fish-like,
with external gills, or,
with legs developing.   ....   To p.119. AMPHIBIANS - 24

1B   Larger; a fish or fish-like. With fins (at least a tail fin). - 2

    2A   With more than one pair of gill openings.
Without paired fins.   Otherwise eel-like. ..... LAMPREYS - 3

    2B   Eel or flatfish.   Only one pair of paired fins. .... - 5

    2C   With one pair of gill-openings and
        two pairs of paired fins.
A typical fish.   ....
- 6

3A   Mouth with lobes (not a sucker). Eyes undeveloped.
Uniform colour all over. No separate dorsal fin.
Small; to about 10cm long.

AMMOCOETE = LARVAL LAMPREY

3B   Mouth surrounded by a sucker. Eyes present.
Over 10cm long. Two separate dorsal fins. ADULT LAMPREY - 4

    4A   In lower reaches of larger rivers.
Over 45cm long.
Colour pattern mottled.

SEA LAMPREY - *Petromyzon marinus*

    4B   In small streams and lake tributaries.
Small; to 15cm long.

BROOK LAMPREY - *Lampetra planeri*

    4C   In large rivers and estuaries.
Up to 40cm long.

RIVER LAMPREY - *Petromyzon fluviatilis*

CHECKLIST*	NOTES
Anguilla anguilla	
Platichthys flesus	
Acipenser sturio	
Alosa alosa	
A. fallax	

GILL-RAKERS - forwardly directed projections from the gill arches. Counts are made on the first arch.

gill-rakers

gill arch

gill filaments

*This list includes all fishes which migrate through estuaries except for lampreys and some salmonids.

5A   Dorsal, caudal (tail) and anal fins continuous.
     Elongate; from 7.5cm to 40cm in males or to 100cm in females.

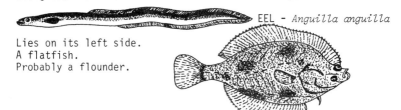

EEL - *Anguilla anguilla*

5B   Lies on its left side.
     A flatfish.
     Probably a flounder.

FLOUNDER - *Platichthys flesus*

6A   Mouth small; tail-fin shark-like.
     With hard bony scales.
     To 400cm (14ft.) long.

STURGEON - *Acipenser sturio*

6B   General appearance herring-like, but deeper bodied.
     One dorsal fin only.
     Fins and scales small.   ....   SHAD - 7

6C   Not a sturgeon or a shad.   .... - 8

7A   Base of anal fin long.   No pattern.
     With about 70 gill-rakers.
     To 60cm long.
     In estuaries in spring.

anal fin

ALLIS SHAD - *Alosa alosa*

7B   Base of anal fin about as long as
        base of dorsal fin.
     Some markings along back.
     With about 30 gill-rakers.
     To 50cm long.
     In estuaries in spring,
        & Killarney Lakes. TWAITE SHAD & KILLARNEY SHAD - *Alosa fallax*

dorsal fin

NOTES

Checklist - see p.108.

PARR - young salmonids after the fry stage.

8A   With an adipose dorsal fin.

                                         .... - 9

8B-D   With two dorsal fins, each with fin rays (B), or with
          a dorsal fin with two obviously different parts (C or D).

B-   C-   D-

                                         To p.111 - 13

8E   With only one dorsal fin.

                                    .....   To p.113 - 16

9A   Uniformly coloured, silvery or dull,
          without definite colour pattern.
          Scales large. ....   - 10

9B   Small; with a series of
          dark "parr" marks
          along the sides.   ....   Juvenile SALMONIDS

9C   With red or dark spots on gill covers,
          adipose dorsal fin, sides etc.
          Scales small.   .... Typical SALMONIDS - 11

10A   Small; to 20cm long, slender and delicate.
        Fresh specimens smell of cucumber.
        About 60 scales along length.
        Estuarine, in spring.

                          SMELT - *Osmerus eperlanus*

10B   Usually small, about 25cm long, but may be much larger.
        Fewer than 100 scale rows from head to tail.
        In shoals in certain lakes.

                      WHITEFISH = POLLAN - *Coregonus*

CHECKLIST   NOTES

*Osmerus eperlanus**

Coregonus sp.*

Salvelinus alpinus

Salmo salar

S.   *trutta*

S.   *gairdnerii*

VOMER - a median bone in the roof   *vomer* –
        of the mouth in fishes.   with &
                                                                without
                                                                teeth

LENGTH OF JAW-BONE (MAXILLA) - short   & long –

*From p.107.

REFERENCES

MAITLAND, P. 1970. The origin and present distribution of
    Coregonus in the British Isles. pp.99-114 *in* C.C. Lindsey
    & C.S. Woods (Eds.) *Biology of Coregonid Fishes*.   Univ.
    of Manitoba Press. Winnipeg. 560pp.
WENT, A.E.J. 1971. The distribution of Irish Char (*Salve-
    linus alpinus*).   *Irish Fish. Investigations*. A (6):5-11.

11A   Teeth present on shaft of vomer bone in mid-line in roof of mouth.
                                         SALMON & TROUT - 12

11B   Teeth absent from shaft of vomer.
       Scales small; about 200 scale rows from head to tail.
       Markings appear light on a dark background.
       May have red spots without "haloes".
       In lakes; spawns in tributaries.

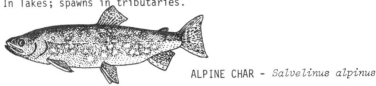

                              ALPINE CHAR - *Salvelinus alpinus*

12A   Jaw bone short - extends back to level of pupil of eye.
       10 - 13 scales above lateral line.
       10 - 12 branched dorsal fin-rays.
       No red colour in adipose dorsal fin.
       Red spots present on parr are
           without "haloes".
       10 - 12 parr marks.
       Tail narrow.
       4 or fewer spots
           on gill covers.   ATLANTIC SALMON - *Salmo salar*

12B   Jaw bone longer - extends to rear of eye
       13 - 16 scales above lateral line.
        8 - 10 branched dorsal fin-rays.
       Adipose dorsal fin red or orange.
       Red spots usually present with
           light-coloured "haloes".
       9 or 10 parr marks in young fish.
       Tail thicker.
       4 or more spots on gill-covers.
                              BROWN TROUT = SEA TROUT - *Salmo trutta*

12C   Dark spots present on caudal (tail) fin.
       Pectoral fin smaller than for Brown Trout.
       Adipose dorsal fin olive with black margin or dark spots.

       pectoral fin   RAINBOW TROUT - *Salmo gairdnerii*

CHECKLIST   NOTES

*Gasterosteus aculeatus**

*Pungitius pungitius**

Pomatoschistus microps

Crenimugil labrosus

*In inland waters also.

13A   With spines (2 or more up to 12) in front of typical dorsal fin.
       Fish small; to 6cm.
       Coastal and inland waters.   ....   STICKLEBACKS - 14

13B   With two webbed dorsal fins - either separate or touching.   - 15

14A   Two, 3 or 4 stout spines on back.

THREE-SPINED STICKLEBACK - *Gasterosteus aculeatus*

14B   About 10 short spines on back.
       More slender build.

TEN-SPINED STICKLEBACK - *Pungitius pungitius*

15A   Dorsal fins almost touching; second fin the larger.
       Small; to 5cm.   Slender.
       Bottom-living in estuaries.

ESTUARINE GOBY - *Pomatoschistus microps*

15B   Dorsal fins spaced far apart.
       Shoals near surface in estuaries.

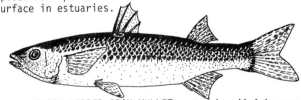

THICK-LIPPED GRAY MULLET - *Crenimugil labrosus*

15C   See next page.   .... - 15C

CHECKLIST*

Perca fluviatilis

Esox lucius

Abramis brama

Cyprinus carpio

Carassius auratus

Gobio gobio *

Nemacheilus barbatula *

Tinca tinca *

Phoxinus phoxinus *

Scardinius erythrophthalmus *

Leuciscus rutilus *

L.   leuciscus *

*This list includes all fishes which are totally adapted to fresh-
water.   See pages 113 - 117.

REFERENCES
BRACKEN, J.J. & M.P. KENNEDY. 1967. A Key to the identification
of the eggs and young stages of coarse fish in Irish waters.
Sci. Proc. Roy. Dublin Soc. Ser. B 2(12):99-108.

15C   Dorsal fins touch;
        first fin the larger.
    Deep-bodied.
    Boldly marked.
    Shoals in lakes and
        their tributaries.

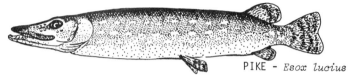

PERCH - *Perca fluviatilis*

16A   Dorsal fin placed far back.
    Jaws large.

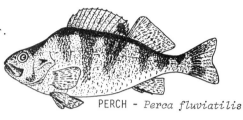

PIKE - *Esox lucius*

16B   Leading edge of dorsal fin in middle third of body. .... - 17

17A   Base of dorsal fin much shorter
        than base of anal fin.
    Deep-bodied.   To 50cm long.
    In lakes and their tributaries.

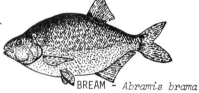

BREAM - *Abramis brama*

17B   Base of dorsal fin much longer than anal fin base.   .... - 18

17C   Bases of dorsal and anal fins similar in length.   .... - 19

18A   With one pair of barbels (whiskers).
    To 60cm long or more.

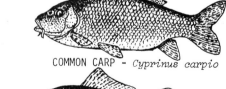

COMMON CARP - *Cyprinus carpio*

18B   Without barbels.
    Brownish, with slight
        golden sheen.

Feral GOLDFISH - *Carassius auratus*

NOTES

Checklist — see p.112.

19A   With one or more pairs of barbels.
       Fish small; to 15cm.   Slender.
       Bottom-living or in shoals.   ...... - 20

                                                 'barbels
19B   Without barbels.
       Small or large.
       Mostly mid-water fishes.   .... - 21

  20A   With one pair of barbels.
         Some stripes on fins.
         Tail-fin forked.
         Usually in shoals.

                                    GUDGEON - *Gobio gobio*

  20B   With three pairs of barbels.
         Irregular mottled colour pattern.
         Tail-fin not forked.
         Secretive; not in shoals.

                          STONE LOACH - *Nemacheilus barbatula*

21A   Tail-fin forked.   .... - 22

21B   Tail-fin not forked.
       Thick heavy fish.

                                    TENCH - *Tinca tinca*

  22A   Small; to 10cm long.
         Dark colour pattern.
         Lives in shoals.

                    BRICKEEN = MINNOW - *Phoxinus phoxinus*

  22B   Larger; no dark pattern; some red colour in some fins.   - 23

NOTES

Checklist – see p.112.

SCALES ABOVE LATERAL LINE –
       count a row of scales from
       the lateral line to the
       front of the dorsal fin.

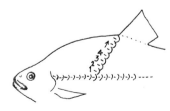

23A Red predominates in all fins except dorsal fin.
     Seven scales above lateral line.
     Common and widespread.
     In lakes etc.

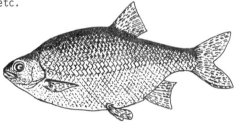

RUDD - *Scardinius erythrophthalmus*

23B Very similar to rudd in all respects, but has
     - less red in fins and eyes.
     - scales smaller; eight scales above lateral line.
     In some lakes etc.

ROACH - *Leuciscus rutilus*

23C More elongate and silvery than roach or rudd.
     Less red in fins.
     Usually in flowing water.

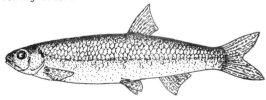

DACE - *Leuciscus leuciscus*

CHECKLIST   NOTES

Rana temporaria

Bufo calamita

Triturus vulgaris

24A   Aquatic larvae = tadpoles.
        Under 3cm long.
        With or without legs or external gills.   ....   - 25

24B   Adult in shape.   Able to hop or crawl out of water.
        Over 2cm long, or if less, then
            without a tail.   ....   - 26

    25A   Black or nearly so.
            Squat; head and trunk short without a neck.
            If only 2 legs are present then
                they are the hind ones.

                                        FROG or TOAD TADPOLES

    25B   Light coloured.
            Elongate.
            Front legs develop faster than hind limbs. .... NEWT TADPOLES

26A   Hind legs much longer than forelegs.
        Hind feet webbed for swimming.
        Skin smooth.
        Colour & pattern variable; usually
            with some dark stripes or marks
            especially on hind limbs.

                            COMMON FROG - *Rana temporaria*

26B   Hind legs & feet as for common frog.
        Skin "rough = warty".
        Golden-yellow stripe down middle of back.
        In some sandy coastal areas.

                    NATTERJACK TOAD - *Bufo calamita*

26C   Fore and hind limbs short and thin.
        Trunk elongate; tail long.
        Skin smooth (without scales).
        Aquatic in spring.

                COMMON NEWT - *Triturus vulgaris* - 27

    27A   Dark back, orange and red underneath with many dark spots.
            Aquatic stage has a crest along the back & tail.   MALE NEWT

    27B   Uniform light brown colour; tail narrower. ...... FEMALE NEWT

IRISH NATIONAL GRID MAPPING

An Irish Biological Records Centre has been established recently to co-ordinate recordings of the distributions of plants and animals. This process has been going on for over 100 years but progressively the techniques for recording have been refined. Most records have been based on 40 vice-counties - West Galway, North Kerry, Antrim, etc., but now the aim is to plot the presence of species on the Irish National Grid system.

This can be found on some Ordnance Survey maps and is the grid reference system used (and explained) in the Illustrated Road Book of Ireland (second Illustrated Edition, revised 1970. Automobile Assn., Dublin. 282pp. + maps). As shown opposite, Ireland is divided into several major squares referred to as B, C, D, etc. The sides of each are 100Km (Kilometres) long (=62 miles).

These major squares are each subdivided into 100 smaller areas which are 10x10Km square. Ireland has close to 1,000 of these and they are referred to by numbers within each major square. The horizontal (ordinate) series of numbers are read first, then those on the vertical axis. So the square for Athlone (a, opposite) is NO6 and for Limerick (b, opposite) R55.

The subdivision of squares is usually taken one step further by again subdividing the axes of the smaller squares into 10 lengths. So we can have M3534 (c, opposite) for a 1x1Km square just south-east of the centre of square M33. Or again C8020 (d, opposite) is a 1x1Km square in the extreme south-west corner of the 10x10Km square C82. These small squares pin-point your record to within about half a mile and avoid the all-too-frequent question "Which River Blackwater does he mean?"

If you wish your identifications to be used in plotting the distributions of the Irish plants and animals then make a note of the grid reference each time you identify a specimen from a new locality, and note the habitat conditions. Your records, whether after a single productive season of field work, or after several years of dabbling, will be very welcome at the Biological Records Centre, An Foras Forbartha, St. Martin's House, Dublin 4. Booklets and advice about the scheme are available from this address. The job is immense and must rely heavily on amateurs, especially since the common species require the most work. Your records may not be used or published at once but the more records there are the sooner someone will work on them.

This scheme will have more than scientific or historic value. It is a base-line for conservation measures to protect the Irish heritage of nature. And conservation is an attempt to sustain you and the rest of mankind in a physically, biologically and socially healthy environment.

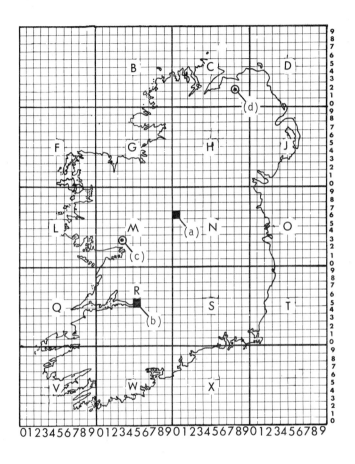

INDEX TO PLANT GENERA

INDEX TO PLANT COMMON NAMES

INDEX TO ANIMAL GENERA (Excluding Insects)

INDEX TO ANIMAL COMMON NAMES (Excluding Insects)

INDEX TO INSECT GENERA

INDEX TO INSECT COMMON NAMES[*]

[*]The words Dun and Spinner for Mayflies are abbreviated to "D" & "S".

COMMON SYNONYMS

The following list gives some alternative scientific names (synonyms) which occur commonly in the literature. Only those synonyms which have been or may become the established name for a species are given, since the only purpose is to facilitate reference to books and papers. All generic names on this page are marked with an asterisk on the index pages.